Free Study Tips DVD

In addition to the tips and content in this guide, we have created a FREE DVD with helpful study tips to further assist your exam preparation. **This FREE Study Tips DVD provides you with top-notch tips to conquer your exam and reach your goals.**

Our simple request in exchange for the strategy-packed DVD is that you email us your feedback about our study guide. We would love to hear what you thought about the guide, and we welcome any and all feedback—positive, negative, or neutral. It is our #1 goal to provide you with top-quality products and customer service.

To receive your **FREE Study Tips DVD**, email freedvd@apexprep.com. Please put "FREE DVD" in the subject line and put the following in the email:

  a. The name of the study guide you purchased.

  b. Your rating of the study guide on a scale of 1-5, with 5 being the highest score.

  c. Any thoughts or feedback about your study guide.

  d. Your first and last name and your mailing address, so we know where to send your free DVD!

Thank you!

SSAT Upper Level Tutor

SSAT Upper Level Prep Books 2020 and 2021 and Practice Test Questions for the Secondary School Admission Test [Includes Detailed Answer Explanations]

APEX Test Prep

Table of Contents

Test Taking Strategies

1. Reading the Whole Question

A popular assumption in Western culture is the idea that we don't have enough time for anything. We speed while driving to work, we want to read an assignment for class as quickly as possible, or we want the line in the supermarket to dwindle faster. However, speeding through such events robs us from being able to thoroughly appreciate and understand what's happening around us. While taking a timed test, the feeling one might have while reading a question is to find the correct answer as quickly as possible. Although pace is important, don't let it deter you from reading the whole question. Test writers know how to subtly change a test question toward the end in various ways, such as adding a negative or changing focus. If the question has a passage, carefully read the whole passage as well before moving on to the questions. This will help you process the information in the passage rather than worrying about the questions you've just read and where to find them. A thorough understanding of the passage or question is an important way for test takers to be able to succeed on an exam.

2. Examining Every Answer Choice

Let's say we're at the market buying apples. The first apple we see on top of the heap may *look* like the best apple, but if we turn it over we can see bruising on the skin. We must examine several apples before deciding which apple is the best. Finding the correct answer choice is like finding the best apple. Although it's tempting to choose an answer that seems correct at first without reading the others, it's important to read each answer choice thoroughly before making a final decision on the answer. The aim of a test writer might be to get as close as possible to the correct answer, so watch out for subtle words that may indicate an answer is incorrect. Once the correct answer choice is selected, read the question again and the answer in response to make sure all your bases are covered.

3. Eliminating Wrong Answer Choices

Sometimes we become paralyzed when we are confronted with too many choices. Which frozen yogurt flavor is the tastiest? Which pair of shoes look the best with this outfit? What type of car will fill my needs as a consumer? If you are unsure of which answer would be the best to choose, it may help to use process of elimination. We use "filtering" all the time on sites such as eBay® or Craigslist® to eliminate the ads that are not right for us. We can do the same thing on an exam. Process of elimination is crossing out the answer choices we know for sure are wrong and leaving the ones that might be correct. It may help to cover up the incorrect answer choice. Covering incorrect choices is a psychological act that alleviates stress due to the brain being exposed to a smaller amount of information. Choosing between two answer choices is much easier than choosing between all of them, and you have a better chance of selecting the correct answer if you have less to focus on.

4. Sticking to the World of the Question

When we are attempting to answer questions, our minds will often wander away from the question and what it is asking. We begin to see answer choices that are true in the real world instead of true in the world of the question. It may be helpful to think of each test question as its own little world. This world may be different from ours. This world may know as a truth that the chicken came before the egg or may assert that two plus two equals five. Remember that, no matter what hypothetical nonsense may be in the question, assume it to be true. If the question states that the chicken came before the egg, then choose your answer based on that truth. Sticking to the world of the question means placing all of our biases and assumptions aside and relying on the question to guide us to the correct answer. If we are simply looking for answers that are correct based on our own judgment, then we may choose incorrectly. Remember an answer that is true does not necessarily answer the question.

5. Key Words

If you come across a complex test question that you have to read over and over again, try pulling out some key words from the question in order to understand what exactly it is asking. Key words may be words that surround the question, such as *main idea, analogous, parallel, resembles, structured,* or *defines*. The question may be asking for the main idea, or it may be asking you to define something. Deconstructing the sentence may also be helpful in making the question simpler before trying to answer it. This means taking the sentence apart and obtaining meaning in pieces, or separating the question from the foundation of the question. For example, let's look at this question:

> Given the author's description of the content of paleontology in the first paragraph, which of the following is most parallel to what it taught?

The question asks which one of the answers most *parallels* the following information: The *description* of paleontology in the first paragraph. The first step would be to see *how* paleontology is described in the first paragraph. Then, we would find an answer choice that parallels that description. The question seems complex at first, but after we deconstruct it, the answer becomes much more attainable.

6. Subtle Negatives

Negative words in question stems will be words such as *not, but, neither,* or *except*. Test writers often use these words in order to trick unsuspecting test takers into selecting the wrong answer—or, at least, to test their reading comprehension of the question. Many exams will feature the negative words in all caps (*which of the following is NOT an example*), but some questions will add the negative word seamlessly into the sentence. The following is an example of a subtle negative used in a question stem:

> According to the passage, which of the following is *not* considered to be an example of paleontology?

If we rush through the exam, we might skip that tiny word, *not*, inside the question, and choose an answer that is opposite of the correct choice. Again, it's important to read the question fully, and double check for any words that may negate the statement in any way.

7. Spotting the Hedges

The word "hedging" refers to language that remains vague or avoids absolute terminology. Absolute terminology consists of words like *always, never, all, every, just, only, none,* and *must*. Hedging refers to words like *seem, tend, might, most, some, sometimes, perhaps, possibly, probability,* and *often*. In some cases, we want to choose answer choices that use hedging and avoid answer choices that use absolute terminology. It's important to pay attention to what subject you are on and adjust your response accordingly.

8. Restating to Understand

Every now and then we come across questions that we don't understand. The language may be too complex, or the question is structured in a way that is meant to confuse the test taker. When you come across a question like this, it may be worth your time to rewrite or restate the question in your own words in order to understand it better. For example, let's look at the following complicated question:

> Which of the following words, if substituted for the word *parochial* in the first paragraph, would LEAST change the meaning of the sentence?

Let's restate the question in order to understand it better. We know that they want the word *parochial* replaced. We also know that this new word would "least" or "not" change the meaning of the sentence. Now let's try the sentence again:

> Which word could we replace with *parochial,* and it would not change the meaning?

Restating it this way, we see that the question is asking for a synonym. Now, let's restate the question so we can answer it better:

> Which word is a synonym for the word *parochial?*

Before we even look at the answer choices, we have a simpler, restated version of a complicated question.

9. Predicting the Answer

After you read the question, try predicting the answer *before* reading the answer choices. By formulating an answer in your mind, you will be less likely to be distracted by any wrong answer choices. Using predictions will also help you feel more confident in the answer choice you select. Once you've chosen your answer, go back and reread the question and answer choices to make sure you have the best fit. If you have no idea what the answer may be for a particular question, forego using this strategy.

10. Avoiding Patterns

One popular myth in grade school relating to standardized testing is that test writers will often put multiple-choice answers in patterns. A runoff example of this kind of thinking is that the most common answer choice is "C," with "B" following close behind. Or, some will advocate certain made-up word patterns that simply do not exist. Test writers do not arrange their correct answer choices in any kind of pattern; their choices are randomized. There may even be times where the correct answer choice will be the same letter for two or three questions in a row, but we have no way of knowing when or if this might happen. Instead of trying to figure out what choice the test writer probably set as being correct, focus on what the *best answer choice* would be out of the answers you are presented with. Use the tips above, general knowledge, and reading comprehension skills in order to best answer the question, rather than looking for patterns that do not exist.

FREE DVD OFFER

Achieving a high score on your exam depends not only on understanding the content, but also on understanding how to apply your knowledge and your command of test taking strategies. **Because your success is our primary goal, we offer a FREE Study Tips DVD. It provides top-notch test taking strategies to help you optimize your testing experience.**

Our simple request in exchange for the strategy-packed DVD is that you email us your feedback about our study guide.

To receive your **FREE Study Tips DVD**, email freedvd@apexprep.com. Please put "FREE DVD" in the subject line and put the following in the email:

a. The name of the study guide you purchased.

b. Your rating of the study guide on a scale of 1-5, with 5 being the highest score.

c. Any thoughts or feedback about your study guide.

d. Your first and last name and your mailing address, so we know where to send your free DVD!

Introduction to the Upper Level SSAT

Function of the Test

The Secondary School Admission Tests (SSATs) are standardized tests intended for applicants to independent or private schools in the United States. There are three levels or iterations of the test based on the grade level of the student. The Upper Level SSAT is taken by students who are currently in grades 8 through 11 and are applying for entrance as a ninth through twelfth grader. Because it is a standardized test, the Upper Level SSAT enables admissions counselors to equitably assess applicants from around the country more so than other factors such as GPA, which can vary based on the rigor of the school, the course load, and grading policies. The test evaluates a candidate's math, reading, and verbal skills, and includes an unscored writing sample, which demonstrates a candidate's writing ability.

Test Administration

The Upper Level SSAT is administered in both a standard testing format and a Flex testing format. The standard format is offered on eight specific Saturdays throughout the year in the United States and in some international locations, while the Flex version is administered to a student on any other date besides one of the eight Saturdays that the standard test is offered. Students with pre-approved accommodations for religious observances may take the standard test on certain Sundays as well.

The Flex test version is an option for students who cannot attend one of the eight standard test dates. It is usually administered by member schools or an educational consultant. Although students may only take the Flex SSAT once per academic year, they can also take the test on any or all of the standard test dates.

Both versions of the Upper Level SSAT are paper and pencil tests. To register for the Upper Level SSAT, interested students must create an account on the SSAT website. From the account, students also print their registration tickets and receive their test scores. Registration for a given standard test date opens about 10 weeks before the date. Late registration begins 3 weeks before the given test administration date and rush registration starts 10 days before the test date. Both late and rush registrations incur additional fees. A student may retake the Upper Level SSAT on any or all of the eight times it is offered throughout the year.

Different testing accommodations are available for students with documents disabilities. Students who require accommodations must apply and submit appropriate documentation. Once they receive approval, they can register for the test. Note that the deadline to request accommodations is generally three weeks prior to the test date. Requesting approval is only required once in an academic year, even if the test taker retakes the test numerous times.

Test Format

The Upper Level SSAT consists of 167 questions, all of which are multiple choice except for the unscored written response that comprises the Writing Sample section. The writing sample produced is sent to the prospective schools a test taker is interested in in order to demonstrate his or her writing ability. In the Writing Sample section, students are offered two prompts—one that should yield a traditional essay and one that will yield a creative story—from which they are to pick one and write an original response.

Sixteen of the 166 multiple-choice questions fall in the final section of the test called the Experimental section. These 6 Verbal, 5 Reading, and 5 Quantitative questions do not impact a test taker's score but are included in order to test their reliability and value for future versions of the SSAT.

The remaining 150 multiple-choice questions on the test are scored and are spread over two Quantitative sections, a Reading section, and a Verbal section. The two Quantitative sections each contain 25 questions covering number concepts, arithmetic, geometry, algebra, and probability. A calculator is not permitted. Each of these sections last 30 minutes.

The 40-minute Reading Comprehension section consists of several reading passages from a variety of domains including literary fiction, social sciences, humanities, and sciences. These passages, which are roughly 250-350 words long, are mostly narratives or arguments. Test takers read the passages and answer 40 multiple-choice questions that assess reading comprehension skills such as identifying the main idea and supporting details, making inferences and drawing conclusions, and identifying the author's purpose or opinion.

The 30-minute Verbal section includes 60 multiple-choice questions: 30 synonyms and 30 analogies. These questions assess a test taker's vocabulary, logical thinking, and verbal reasoning.

The sections of the test are broken down as follows:

Section	Number of Questions	Time Allotted
Writing Sample	1 (unscored)	25 minutes
Break		5 minutes
Quantitative 1	25	30 minutes
Reading	40	40 minutes
Break		10 minutes
Verbal	60	30 minutes
Quantitative 2	25	30 minutes
Experimental	16 (unscored)	15 minutes
Total	167 (150 scored)	3 hours, 5 minutes

Scoring

For each of the 150 scored multiple-choice questions on the Upper Level SSAT, test takers receive one point for each correct answer, or lose a quarter of a point for each incorrect answer. Therefore, it makes sense for test takers to weigh how confident they are in their responses to some degree.

Approximately two weeks after the administration of the Upper Level SSAT, a free scoring report is available online through the student's SSAT account. The score report includes a raw score, a scaled score, a percentile rank, and the total scaled score along with an explanation of the scores. Scaled scores for each section range from 440–710. The total scaled score ranges from 1320–2130, with a mid-point of 1725. Percentile rank scores range from 1–99. Writing section reports can be purchased in a separate score report.

Like all SSAT exams, the Upper Level SSAT is a norm-referenced test. Therefore, a test taker's score is compared to the scores from a norm group of test takers from the prior three years. Test takers are able to see the average norm score for each section of the test on their score report, so they can compare their performance relative to the norm group. Additionally, the percentile rank shows how the test taker performed relative to the norm group. A percentile rank of 90, for example, means the test taker scored better than 90% of the norm group.

There is no universal passing score because different schools set their own expectations and may rely more or less on the test among the various factors that go into an application and admissions decision.

Quantitative

Number and Quantity

Addition, Subtraction, Multiplication, and Division with Whole Numbers

The four basic operations include addition, subtraction, multiplication, and division. The result of addition is a **sum**, the result of subtraction is a **difference**, the result of multiplication is a **product**, and the result of division is a **quotient**. Each type of operation can be used when working with rational numbers; however, the basic operations need to be understood first while using simpler numbers before working with fractions and decimals.

Performing these operations should first be learned using whole numbers. Addition needs to be done column by column. To add two whole numbers, add the ones column first, then the tens columns, then the hundreds, etc. If the sum of any column is greater than 9, a one must be carried over to the next column. For example, the following is the result of 482 + 924:

$$
\begin{array}{r}
{}^{1} \\
482 \\
+924 \\
\hline
1406
\end{array}
$$

Notice that the sum of the ten's column was 10, so a one was carried over to the hundred's column. Subtraction is also performed column by column. Subtraction is performed in the one's column first, then the tens, etc. If the number on top is smaller than the number below, a one must be borrowed from the column to the left. For example, the following is the result of 5,424 − 756:

$$
\begin{array}{r}
4\ \ 13\ \ 11\ \ 14 \\
\cancel{5}\ \ \cancel{4}\ \ \cancel{2}\ \ \cancel{4} \\
-\ \ 7\ \ \ 5\ \ \ 6 \\
\hline
4\ \ \ 6\ \ \ 6\ \ \ 8
\end{array}
$$

Notice that a one is borrowed from the tens, hundreds, and thousands place. After subtraction, the answer can be checked through addition. A check of this problem would be to show that 756 + 4,668 = 5,424.

Multiplication of two whole numbers is performed by writing one on top of the other. The number on top is known as the **multiplicand,** and the number below is the **multiplier.** Perform the multiplication by multiplying the multiplicand by each digit of the multiplier. Make sure to place the ones value of each result under the multiplying digit in the multiplier. Each value to the right is then a 0. The product is found by adding each partial product. For example, the following is the process of multiplying 46 times 37, where 46 is the multiplicand and 37 is the multiplier:

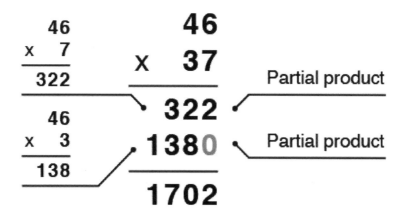

Finally, division can be performed using long division. When dividing a number by another number, the first number is known as the **dividend,** and the second is the **divisor.** For example, with $a \div b = c$, a is the dividend, b is the divisor, and c is the quotient. For long division, place the dividend within the division symbol and the divisor on the outside. For example, with $8{,}764 \div 4$, refer to the first problem in the diagram below. First, there are two 4's in the first digit, 8. This number 2 gets written above the 8. Then, multiply 4 times 2 to get 8, and that product goes below the 8. Subtract to get 8, and then carry down the 7. Continue the same steps. $7 \div 4 = 1$ R3, so 1 is written above the 7. Multiply 4 times 1 to get 4, and write it below the 7. Subtract to get 3, and carry the 6 down next to the 3. Resulting steps give a 9 and a 1. The final subtraction results in a 0, which means that 8,764 is divisible by 4. There are no remaining numbers.

The second example in the image that follows shows that $4{,}536 \div 216 = 21$. The steps are a little different because 216 cannot be contained in 4 or 5, so the first step is placing a 2 above the 3 because there are 2 216's in 453. Finally, the third example shows that $546 \div 31 = 17\,R19$. The 19 is a remainder. Notice that the final subtraction does not result in a 0, which means that 546 is not divisible by 31. The remainder can also be written as a fraction over the divisor to say that $546 \div 31 = 17\frac{19}{31}$.

```
        2191              21               17 r 19
    4 | 8764       216 | 4536        31 | 546
        8↓↓↓             432↓              31↓
        07               216              236
         4↓              216              217
         36                0               19
         36↓
          04
           4
           0
```

If a division problem relates to a real-world application, and a remainder does exist, it can have meaning. For example, consider the third example, $546 \div 31 = 17R19$. Let's say that we had \$546 to spend on calculators that cost \$31 each, and we wanted to know how many we could buy. The division problem would answer this question. The result states that 17 calculators could be purchased, with \$19 left over. Notice that the remainder will never be greater than or equal to the divisor.

Addition, Subtraction, Multiplication, and Division with Positive and Negative Numbers

Once the operations are understood with whole numbers, they can be used with integers. There are many rules surrounding operations with negative numbers. First, consider addition with integers. The sum of two numbers can first be shown using a number line. For example, to add $1 + (-4)$, plot the point 1 on the number line. Then, because a negative number is being added, move 4 units to the left. This process results in landing on -3 on the number line, which is the sum of 1 and -4. If adding a positive number, move to the right.

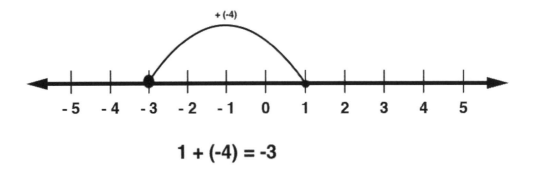

$$1 + (-4) = -3$$

Visualizing this process using a number line is useful for understanding; however, it is not efficient. A quicker process is to learn the rules. When adding two numbers with the same sign, add the absolute values of both numbers, and use the common sign of both numbers as the sign of the sum. For example, to add $-5 + (-6)$, add their absolute values $5 + 6 = 11$. Then, introduce a negative number because both addends are negative. The result is -11. To add two integers with unlike signs, subtract the lesser absolute value from the greater absolute value, and apply the sign of the number with the greater absolute value to the result. For example, the sum $-7 + 4$ can be computed by finding the difference $7 - 4 = 3$ and then applying a negative sign because the value with the larger absolute value is negative. The result is -3. Similarly, the sum $-4 + 7$ can be found by computing the same difference but leaving it as a positive result because the addend with the larger absolute value is positive. Also, recall that any number plus 0 equals that number. This is known as the **Addition Property of 0.**

Subtracting two integers can be computed by changing to addition to avoid confusion. The rule is to add the first number to the opposite of the second number. The opposite of a number is the number on the other side of 0 on the number line, which is the same number of units away from 0. For example, -2 and 2 are opposites. Consider $4 - 8$. Change this to adding the opposite as follows: $4 + (-8)$. Then, follow the rules of addition of integers to obtain -4. Secondly, consider $-8 - (-2)$. Change this problem to adding the opposite as $-8 + 2$, which equals -6. Notice that subtracting a negative number is really adding a positive number.

Multiplication and division of integers are actually less confusing than addition and subtraction because the rules are simpler to understand. If two factors in a multiplication problem have the same sign, the result is positive. If one factor is positive and one factor is negative, the result, known as the **product,** is negative. For example, $(-9)(-3) = 27$ and $9(-3) = -27$. Also, a number times 0 always results in 0. If a problem consists of more than a single multiplication, the result is negative if it contains an odd number of negative factors, and the result is positive if it contains an even number of negative factors. For example, $(-1)(-1)(-1)(-1) = 1$ and $(-1)(-1)(-1)(-1)(-1) = -1$. These two examples of multiplication also bring up another concept. Both are examples of repeated multiplication, which can be written in a more compact notation using exponents. The first example can be written as $(-1)^4 = 1$, and the second example can be written as $(-1)^5 = -1$. Both are exponential expressions: -1 is the **base** in both instances, and 4 and 5 are the respective **exponents.** Note that a negative number raised to an odd power is always negative, and a negative number raised to an even power is always positive. Also, $(-1)^4$ is not the same as -1^4. In the first expression, the negative is included in the parentheses, but it is not in the second expression. The second expression is found by evaluating 1^4 first to get 1 and then applying the negative sign to obtain -1.

A similar theory applies within division. First, consider some vocabulary. When dividing 14 by 2, it can be written in the following ways: $14 \div 2 = 7$ or $\frac{14}{2} = 7$. 14 is the **dividend,** 2 is the **divisor,** and 7 is the **quotient.** If two numbers in a division problem have the same sign, the quotient is positive. If two numbers in a division problem have different signs, the quotient is negative. For example:

$$14 \div (-2) = -7, \text{ and } -14 \div (-2) = 7$$

To check division, multiply the quotient times the divisor to obtain the dividend. Also, remember that 0 divided by any number is equal to 0. However, any number divided by 0 is undefined. It just does not make sense to divide a number by 0 parts.

If more than one operation is to be completed in a problem, follow the **Order of Operations**. The mnemonic device, PEMDAS, for the order of operations states the order in which addition, subtraction, multiplication, and division need to be done. It also includes when to evaluate operations within grouping symbols and when to incorporate exponents. PEMDAS, which some remember by thinking "please excuse my dear Aunt Sally," refers to parentheses, exponents, multiplication, division, addition, and subtraction. First, within an expression, complete any operation that is within parentheses, or any other grouping symbol like brackets, braces, or absolute value symbols. Note that this does not refer to the case when parentheses are used to represent multiplication like $(2)(5)$, wherein an operation is not within parentheses like it is in $(2 \cdot 5)$. Then, any exponents must be computed. Next, multiplication and division are performed from left to right. Finally, addition and subtraction are performed from left to right.

The following is an example in which the operations within the parentheses need to be performed first, so the order of operations must be applied to the exponent, subtraction, addition, and multiplication within the grouping symbol:

$$9 - 3(3^2 - 3 + 4 \cdot 3)$$

$$9 - 3(3^2 - 3 + 4 \cdot 3) \quad \text{Work within the parentheses first}$$

$$= 9 - 3(9 - 3 + 12)$$

$$= 9 - 3(18)$$

$$= 9 - 54$$

$$= -45$$

Addition, Subtraction, Multiplication, and Division with Fractions, Decimals, & Percentages

Operations with Fractions
Once the rules for integers are understood, operations with fractions and decimals can be mastered. Recall that a **rational number** can be written as a fraction and can be converted to a decimal through division. If a rational number is negative, the rules for adding, subtracting, multiplying, and dividing integers must be used. If a rational number is in fractional form, performing addition, subtraction, multiplication, and division is more complicated than when working with integers. First, consider addition. To add two fractions having the same denominator, add the numerators and then reduce the fraction. When an answer is a fraction, it should always be in lowest terms. **Lowest terms** means that every common factor, other than 1, between the numerator and denominator is divided out. For example:

$$\frac{2}{8} + \frac{4}{8} = \frac{6}{8} = \frac{6 \div 2}{8 \div 2} = \frac{3}{4}$$

Both the numerator and denominator of $\frac{6}{8}$ have a common factor of 2, so 2 is divided out of each number to put the fraction in lowest terms. If denominators are different in an addition problem, the fractions must be converted to have common denominators. The **least common denominator (LCD)** of all the given denominators must be found, and this value is equal to the **least common multiple (LCM)** of the denominators. This non-zero value is the smallest number that is a multiple of both denominators. Then, each original fraction can be written as an equivalent fraction using the new denominator. Once in this form, process of adding with like denominators is completed. For example, consider $\frac{1}{3} + \frac{4}{9}$. The LCD is 9 because 9 is the smallest multiple of both 3 and 9. The fraction $\frac{1}{3}$ must be rewritten with 9 as its denominator. Therefore, multiply both the numerator and denominator by 3. Multiplying by $\frac{3}{3}$ is the same as multiplying times 1, so it does not change the value of the fraction. Therefore, an equivalent fraction is $\frac{3}{9}$, and $\frac{1}{3} + \frac{4}{9} = \frac{3}{9} + \frac{4}{9} = \frac{7}{9}$, which is in lowest terms. Subtraction is performed in a similar manner; once the denominators are equal, the numerators are then subtracted. The following is an example of addition of a positive and a negative fraction:

$$-\frac{5}{12} + \frac{5}{9} = -\frac{5 \times 3}{12 \times 3} + \frac{5 \times 4}{9 \times 4}$$

$$-\frac{15}{36} + \frac{20}{36} = \frac{5}{36}$$

Common denominators are not necessary when multiplying and dividing fractions. To multiply two fractions, multiply the numerators together and the denominators together. Then, write the result in lowest terms.

For example:

$$\frac{2}{3} \times \frac{9}{4} = \frac{18}{12} = \frac{3}{2}$$

Alternatively, the fractions could be factored first to cancel out any common factors before performing the multiplication. For example:

$$\frac{2}{3} \times \frac{9}{4} = \frac{2}{3} \times \frac{3 \times 3}{2 \times 2} = \frac{3}{2}$$

This second approach is helpful when working with larger numbers, as common factors might not be obvious. Multiplication and division of fractions are related because the division of two fractions is changed into a multiplication problem. Dividing one fraction by another is equivalent to multiplying the first by the **reciprocal** of the second fraction, so that second fraction must be inverted, or "flipped," to be in reciprocal form. For example:

$$\frac{11}{15} \div \frac{3}{5} = \frac{11}{15} \times \frac{5}{3} = \frac{55}{45} = \frac{11}{9}$$

The fraction $\frac{5}{3}$ is the reciprocal of $\frac{3}{5}$. It is possible to multiply and divide numbers containing a mix of integers and fractions. In this case, convert the integer to a fraction by placing it over a denominator of 1. For example, a division problem involving an integer and a fraction is:

$$3 \div \frac{1}{2} = \frac{3}{1} \times \frac{2}{1} = \frac{6}{1} = 6$$

Finally, when performing operations with rational numbers that are negative, the same rules apply as when performing operations with integers. For example, a negative fraction times a negative fraction results in a positive value, and a negative fraction subtracted from a negative fraction results in a negative value.

Operations with Decimals

Operations can be performed on rational numbers in decimal form. Recall that to write a fraction as an equivalent decimal expression, divide the numerator by the denominator. For example, $\frac{1}{8} = 1 \div 8 = 0.125$.

When working with decimals, it is important to keep track of place value. To add decimals, make sure the decimal places are in alignment so that the numbers are lined up with their decimal points. Then add vertically. If the numbers do not line up because there are extra or missing place values in one of the numbers, then zeros may be used as placeholders. For example, $0.123 + 0.23$ becomes:

$$\begin{array}{r} 0.123 \\ \underline{0.230} \\ 0.353 \end{array}$$

Subtraction is done the same way. Multiplication and division are more complicated. To multiply two decimals, place one on top of the other as in a regular multiplication process and do not worry about lining up the decimal points. Then, multiply as with whole numbers, ignoring the decimals. Finally, in the solution, insert the decimal point as many places to the left as there are total decimal values in the original problem. Here is an example of a decimal multiplication:

$$\begin{array}{rl} 0.52 & \textit{2 decimal places} \\ \underline{\times \quad 0.2} & \textit{1 decimal place} \\ 0.104 & \textit{3 decimal places} \end{array}$$

The answer to 52 times 2 is 104, and because there are three decimal values in the problem, the decimal point is positioned three units to the left in the answer.

The decimal point plays an integral role throughout the whole problem when dividing with decimals. First, set up the problem in a long division format. If the divisor is not an integer, the decimal must be moved to the right as many units as needed to make it an integer. The decimal in the dividend must be moved to the right the same number of places to maintain equality. Then, division is completed normally. Here is an example of long division with decimals:

Long division
with decimals

$$0.06 \overline{\smash{)}12.72}$$

$$\begin{array}{r} 212 \\ 6 \overline{\smash{)}1272} \\ \underline{12} \\ 07 \\ \underline{6} \\ 12 \end{array}$$

Because the decimal point is moved two units to the right in the divisor of 0.06 to turn it into the integer 6, it is also moved two units to the right in the dividend of 12.72 to make it 1,272. The result is 212, and remember that a division problem can always be checked by multiplying the answer by the divisor to see if the result is equal to the dividend.

14

Sometimes it is helpful to round answers that are in decimal form. First, find the place to which the rounding needs to be done. Then, look at the digit to the right of it. If that digit is 4 or less, the number in the place value to its left stays the same, and everything to its right becomes a 0. This process is known as **rounding down**. If that digit is 5 or higher, round up by increasing the place value to its left by 1, and every number to its right becomes a 0. If those 0's are in decimals, they can be dropped. For example, 0.145 rounded to the nearest hundredths place would be rounded up to 0.15 because there is a 5 in the thousandths place, and 0.145 rounded to the nearest tenths place would be rounded down to 0.1 because there is a 4 in the hundredths place and 4 is less than 5.

Converting Non-Negative Fractions, Decimals, and Percentages
Within the number system, different forms of numbers can be used. It is important to be able to recognize each type, as well as work with, and convert between, the given forms. The **real number system** comprises natural numbers, whole numbers, integers, rational numbers, and irrational numbers. Natural numbers, whole numbers, integers, and irrational numbers typically are not represented as fractions, decimals, or percentages. Rational numbers, however, can be represented as any of these three forms. A **rational number** is a number that can be written in the form $\frac{a}{b}$, where a and b are integers, and b is not equal to zero. In other words, rational numbers can be written in a fraction form. The value a is the **numerator,** and b is the **denominator.**

If the numerator is equal to zero, the entire fraction is equal to zero. Non-negative fractions can be less than 1, equal to 1, or greater than 1. Fractions are less than 1 if the numerator is smaller (less than) than the denominator. For example, $\frac{3}{4}$ is less than 1. A fraction is equal to 1 if the numerator is equal to the denominator. For instance, $\frac{4}{4}$ is equal to 1. Finally, a fraction is greater than 1 if the numerator is greater than the denominator: the fraction $\frac{11}{4}$ is greater than 1. When the numerator is greater than the denominator, the fraction is called an **improper fraction**. An improper fraction can be converted to a **mixed number**, which is a combination of both a whole number and a fraction. To convert an improper fraction to a mixed number, divide the numerator by the denominator. Write down the whole number portion, and then write any remainder over the original denominator. For example, $\frac{11}{4}$ is equivalent to $2\frac{3}{4}$. Conversely, a mixed number can be converted to an improper fraction by multiplying the denominator by the whole number and adding that result to the numerator.

Fractions can be converted to decimals. With a calculator, a fraction is converted to a decimal by dividing the numerator by the denominator. For example:

$$\frac{2}{5} = 2 \div 5 = 0.4$$

Sometimes, rounding might be necessary. Consider:

$$\frac{2}{7} = 2 \div 7 = 0.28571429$$

This decimal could be rounded for ease of use, and if it needed to be rounded to the nearest thousandth, the result would be 0.286. If a calculator is not available, a fraction can be converted to a decimal manually. First, find a number that, when multiplied by the denominator, has a value equal to 10, 100, 1,000, etc. Then, multiply both the numerator and denominator by that number. The decimal form of the fraction is equal to the new numerator with a decimal point placed as many place values to the left as there are zeros in the denominator. For example, to convert $\frac{3}{5}$ to a decimal, multiply both the numerator

15

and denominator by 2, which results in $\frac{6}{10}$. The decimal is equal to 0.6 because there is one zero in the denominator, and so the decimal place in the numerator is moved one unit to the left.

In the case where rounding would be necessary while working without a calculator, an approximation must be found. A number close to 10, 100, 1,000, etc. can be used. For example, to convert $\frac{1}{3}$ to a decimal, the numerator and denominator can be multiplied by 33 to turn the denominator into approximately 100, which makes for an easier conversion to the equivalent decimal. This process results in $\frac{33}{99}$ and an approximate decimal of 0.33. Once in decimal form, the number can be converted to a percentage. Multiply the decimal by 100 and then place a percent sign after the number. For example, 0.614 is equal to 61.4%. In other words, move the decimal place two units to the right and add the percentage symbol.

Factorization

Factorization is the process of breaking up a mathematical quantity, such as a number or polynomial, into a product of two or more factors. For example, a factorization of the number 16 is $16 = 8 \times 2$. If multiplied out, the factorization results in the original number. A **prime factorization** is a specific factorization when the number is factored completely using prime numbers only. For example, the prime factorization of 16 is $16 = 2 \times 2 \times 2 \times 2$. A factor tree can be used to find the prime factorization of any number. Within a factor tree, pairs of factors are found until no other factors can be used, as in the following factor tree of the number 84:

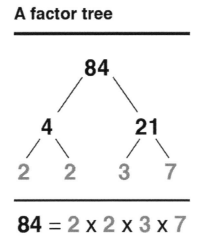

A factor tree

$$84 = 2 \times 2 \times 3 \times 7$$

It first breaks 84 into 21×4, which is not a prime factorization because both of these numbers have other factors. Then, both 21 and 4 are factored into their primes. The final numbers on each branch consist of the numbers within the prime factorization. Therefore, $84 = 2 \times 2 \times 3 \times 7$. Factorization can be helpful in finding greatest common divisors and least common denominators.

Also, a factorization of an algebraic expression can be found. Throughout the process, a more complicated expression can be decomposed into products of simpler expressions. To factor a polynomial, first determine if there is a greatest common factor. If there is, factor it out. For example, $2x^2 + 8x$ has a greatest common factor of $2x$ and can be written as $2x(x + 4)$. Once the greatest common monomial factor is factored out, if applicable, count the number of terms in the polynomial. If there are two terms, is it a difference of squares, a sum of cubes, or a difference of cubes?

If so, the following rules can be used:

$$a^2 - b^2 = (a + b)(a - b)$$

$$a^3 + b^3 = (a + b)(a^2 - ab + b^2)$$

$$a^3 - b^3 = (a + b)(a^2 + ab + b^2)$$

If there are three terms, and if the trinomial is a perfect square trinomial, it can be factored into the following:

$$a^2 + 2ab + b^2 = (a + b)^2$$

$$a^2 - 2ab + b^2 = (a - b)^2$$

If not, try factoring into a product of two binomials by trial and error into a form of $(x + p)(x + q)$. For example, to factor $x^2 + 6x + 8$, determine what two numbers have a product of 8 and a sum of 6. Those numbers are 4 and 2, so the trinomial factors into $(x + 2)(x + 4)$.

Finally, if there are four terms, try factoring by grouping. First, group terms together that have a common monomial factor. Then, factor out the common monomial factor from the first two terms. Next, look to see if a common factor can be factored out of the second set of two terms that results in a common binomial factor. Finally, factor out the common binomial factor of each expression, for example:

$$xy - x + 5y - 5 = x(y - 1) + 5(y - 1)$$

$$(y - 1)(x + 5)$$

After the expression is completely factored, check to see if the factorization is correct by multiplying to try to obtain the original expression. Factorizations are helpful in solving equations that consist of a polynomial set equal to 0. If the product of two algebraic expressions equals 0, then at least one of the factors is equal to 0. Therefore, factor the polynomial within the equation, set each factor equal to 0, and solve. For example, $x^2 + 7x - 18 = 0$ can be solved by factoring into $(x + 9)(x - 2) = 0$. Set each factor equal to 0, and solve to obtain $x = -9$ and $x = 2$.

Solving Practical Math Problems

Solving Problems
One-step problems take only one mathematical step to solve. For example, solving the equation $5x = 45$ is a one-step problem because the one step of dividing both sides of the equation by 5 is the only step necessary to obtain the solution $x = 9$. The **multiplication principle of equality** is the one step used to isolate the variable. The equation is of the form $ax = b$, where a and b are rational numbers. Similarly, the **addition principle of equality** could be the one step needed to solve a problem. In this case, the equation would be of the form $x + a = b$ or $x - a = b$, for real numbers a and b.

A multi-step problem involves more than one step to find the solution, or it could consist of solving more than one equation. An equation that involves both the addition principle and the multiplication principle is a two-step problem, and an example of such an equation is $2x - 4 = 5$. Solving involves adding 4 to both sides and then dividing both sides by 2. An example of a two-step problem involving two separate equations is $y = 3x, 2x + y = 4$. The two equations form a system of two equations that must be solved together in two variables. The system can be solved by the substitution method. Since y is already solved for in terms of x, plug $3x$ in for y into the equation $2x + y = 4$, resulting in $2x + 3x = 4$. Therefore, $5x = 4$

and $x = \frac{4}{5}$. Because there are two variables, the solution consists of a value for both x and for y. Substitute $x = \frac{4}{5}$ into either original equation to find y. The easiest choice is $y = 3x$. Therefore:

$$y = 3 \times \frac{4}{5} = \frac{12}{5}$$

The solution can be written as the ordered pair $\left(\frac{4}{5}, \frac{12}{5}\right)$.

Real-world problems can be translated into both one-step and multi-step problems. In either case, the word problem must be translated from the verbal form into mathematical expressions and equations that can be solved using algebra. An example of a one-step real-world problem is the following: A cat weighs half as much as a dog living in the same house. If the dog weighs 14.5 pounds, how much does the cat weigh? To solve this problem, an equation can be used. In any word problem, the first step must be defining variables that represent the unknown quantities. For this problem, let x be equal to the unknown weight of the cat. Because two times the weight of the cat equals 14.5 pounds, the equation to be solved is: $2x = 14.5$. Use the multiplication principle to divide both sides by 2. Therefore, $x = 7.25$. The cat weighs 7.25 pounds.

Most of the time, real-world problems are more difficult than this one and are multi-step problems. The following is an example of a multi-step problem: The sum of two consecutive page numbers is equal to 437. What are those page numbers? First, define the unknown quantities. If x is equal to the first page number, then $x + 1$ is equal to the next page number because they are consecutive integers. Their sum is equal to 437, and this statement translates to the equation $x + (x + 1) = 437$. To solve, first collect like terms to obtain $2x + 1 = 437$.

Then, subtract 1 from both sides and then divide by 2. The solution to the equation is $x = 218$. Therefore, the two consecutive page numbers that satisfy the problem are 218 and 219. It is always important to make sure that answers to real-world problems make sense. For instance, if the solution to this same problem resulted in decimals, that should be a red flag indicating the need to check the work. Page numbers are whole numbers; therefore, if decimals are found to be answers, the solution process should be double-checked to see where mistakes were made.

In general, to solve problems, follow these steps: Identify the variables that are known, decide which equation should be used, substitute the numbers, and solve. To solve an equation for the amount of time that has elapsed since an event, use the equation $T = L - E$ where T represents the elapsed time, L represents the later time, and E represents the earlier time. For example, the Minnesota Vikings have not appeared in the Super Bowl since 1976. If the year is now 2017, how long has it been since the Vikings were in the Super Bowl? The later time, L, is 2017, E = 1976 and the unknown is T. Substituting these numbers, the equation is T = 2017 − 1976, and so T = 41. It has been 41 years since the Vikings have appeared in the Super Bowl. Questions involving total cost can be solved using the formula, C = I + T where C represents the total cost, I represents the cost of the item purchased, and T represents the tax amount. To find the length of a rectangle given the area = 32 square inches and width = 8 inches, the formula A = L × W can be used.

Substitute 32 for A and substitute 8 for w, giving the equation 32 = L × 8. This equation is solved by dividing both sides by 8 to find that the length of the rectangle is 4. The formula for volume of a rectangular prism is given by the equation V = L × W × H. If the length of a rectangular juice box is 4 centimeters, the width is 2 centimeters, and the height is 8 centimeters, what is the volume of this box? Substituting in the formula we find V = 4 × 2 × 8, so the volume is 64 cubic centimeters. In a similar fashion as those previously shown, the mass of an object can be calculated given the formula, Mass = Density × Volume.

Solving Single- and Multistep Problems Involving Percentages

Percentages are defined to be parts per one hundred. To convert a decimal to a percentage, move the decimal point two units to the right and place the percent sign after the number. Percentages appear in many scenarios in the real world. It is important to make sure the statement containing the percentage is translated to a correct mathematical expression. Be aware that it is extremely common to make a mistake when working with percentages within word problems.

An example of a word problem containing a percentage is the following: 35% of people speed when driving to work. In a group of 5,600 commuters, how many would be expected to speed on the way to their place of employment? The answer to this problem is found by finding 35% of 5,600. First, change the percentage to the decimal 0.35. Then compute the product: $0.35 \times 5,600 = 1,960$. Therefore, it would be expected that 1,960 of those commuters would speed on their way to work based on the data given. In this situation, the word "of" signals to use multiplication to find the answer. Another way percentage is used is in the following problem: *Teachers work 8 months out of the year. What percent of the year do they work?* To answer this problem, find what percent of 12 the number 8 is, because there are 12 months in a year. Therefore, divide 8 by 12, and convert that number to a percentage:

$$\frac{8}{12} = \frac{2}{3} = 0.66\overline{6}$$

The percentage rounded to the nearest tenth place tells us that teachers work 66.7% of the year. Percentages also appear in real-world application problems involving finding missing quantities like in the following question: 60% of what number is 75? To find the missing quantity, an equation can be used. Let x be equal to the missing quantity. Therefore, $0.60x = 75$. Divide each side by 0.60 to obtain 125. Therefore, 60% of 125 is equal to 75.

Sales tax is an important application relating to percentages because tax rates are usually given as percentages. For example, a city might have an 8% sales tax rate. Therefore, when an item is purchased with that tax rate, the real cost to the customer is 1.08 times the price in the store. This is the same as adding the 0.08 times the cost and adding it to the original cost but it saves a step. For example, a $25 pair of jeans costs the customer $25 \times 1.08 = \$27$. Sales tax rates can also be determined if they are unknown when an item is purchased. If a customer visits a store and purchases an item for $21.44, but the price in the store was $19, they can find the tax rate by first subtracting $21.44 − $19 to obtain $2.44, the sales tax amount. The sales tax is a percentage of the in-store price. Therefore, the tax rate is $\frac{2.44}{19} = 0.128$, which has been rounded to the nearest thousandths place. In this scenario, the actual sales tax rate given as a percentage is 12.8%.

Fraction and Percent Equivalencies

Fractions appear in everyday situations, and in many scenarios, they appear in the real-world as ratios and in proportions. A *ratio* is formed when two different quantities are compared. For example, in a group of 50 people, if there are 33 females and 17 males, the ratio of females to males is 33 to 17. This expression can be written in the fraction form, $\frac{33}{17}$, or by using the ratio symbol, 33:17. The order of the number matters when forming ratios. In the same setting, the ratio of males to females is 17 to 33, which is equivalent to $\frac{17}{33}$ or 17:33. A *proportion* is an equation involving two ratios. The equation $\frac{a}{b} = \frac{c}{d}$, or $a:b = c:d$ is a proportion, for real numbers a, b, c, and d. Usually, in one ratio, one of the quantities is unknown, and cross-multiplication is used to solve for the unknown. Consider $\frac{1}{4} = \frac{x}{5}$. To solve for x, cross-multiply to obtain $5 = 4x$. Divide each side by 4 to obtain the solution $x = \frac{5}{4}$. It is also true that percentages are ratios in which the second term is 100. For example, 65% is 65:100 or $\frac{65}{100}$. Therefore, when working with percentages, one is also working with ratios.

Solving Real-World Problems Involving Proportions

Fractions appear in everyday situations, and in many scenarios, they appear in the real-world as ratios and in proportions. A **ratio** is formed when two different quantities are compared. For example, in a group of 50 people, if there are 33 females and 17 males, the ratio of females to males is 33 to 17. This expression can be written in the fraction form, $\frac{33}{17}$, or by using the ratio symbol, 33:17. The order of the number matters when forming ratios. In the same setting, the ratio of males to females is 17 to 33, which is equivalent to $\frac{17}{33}$ or 17:33. A **proportion** is an equation involving two ratios. The equation $\frac{a}{b} = \frac{c}{d}$, or $a:b = c:d$ is a proportion, for real numbers a, b, c, and d. Usually, in one ratio, one of the quantities is unknown, and cross-multiplication is used to solve for the unknown. Consider $\frac{1}{4} = \frac{x}{5}$. To solve for x, cross-multiply to obtain $5 = 4x$. Divide each side by 4 to obtain the solution $x = \frac{5}{4}$. It is also true that percentages are ratios in which the second term is 100. For example, 65% is 65:100 or $\frac{65}{100}$. Therefore, when working with percentages, one is also working with ratios.

Real-world problems frequently involve proportions. For example, consider the following problem: If 2 out of 50 pizzas are usually delivered late from a local Italian restaurant, how many would be late out of 235 orders? The following proportion would be solved with x as the unknown quantity of late pizzas: $\frac{2}{50} = \frac{x}{235}$. Cross multiplying results in $470 = 50x$. Divide both sides by 50 to obtain $x = \frac{470}{50}$, which, in lowest terms, is equal to $\frac{47}{5}$. In decimal form, this improper fraction is equal to 9.4. Because it does not make sense to answer this question with decimals (portions of pizzas do not get delivered) the answer must be rounded. Traditional rounding rules would say that 9 pizzas would be expected to be delivered late. However, to be safe, rounding up to 10 pizzas out of 235 would probably make more sense.

Solving Real-World Problems Involving Ratios and Rates of Change

Recall that a **ratio** is the comparison of two different quantities. Comparing 2 apples to 3 oranges results in the ratio 2:3, which can be expressed as the fraction $\frac{2}{3}$. Note that order is important when discussing ratios. The number mentioned first is the numerator, and the number mentioned second is the denominator. The ratio 2:3 does not mean the same quantity as the ratio 3:2. Also, it is important to make sure than when discussing ratios that have units attached to them, the two quantities use the same units. For example, to think of 8 feet to 4 yards, it would make sense to convert 4 yards to feet by multiplying by 3. Therefore, the ratio would be 8 feet to 12 feet, which can be expressed as the fraction $\frac{8}{12}$. Also, note that it is proper to refer to ratios in lowest terms. Therefore, the ratio of 8 feet to 4 yards is equivalent to the fraction $\frac{2}{3}$.

Many real-world problems involve ratios. Often, problems with ratios involve proportions, as when two ratios are set equal to find the missing amount. However, some problems involve deciphering single ratios. For example, consider an amusement park that sold 345 tickets last Saturday. If 145 tickets were sold to adults and the rest of the tickets were sold to children, what would the ratio of the number of adult tickets to children's tickets be? A common mistake would be to say the ratio is 145:345. However, 345 is the total number of tickets sold, not the number of children's tickets. There were $345 - 145 = 200$ tickets sold to children. The correct ratio of adult to children's tickets is 145:200. As a fraction, this expression is written as $\frac{145}{200}$, which can be reduced to $\frac{29}{40}$.

While a ratio compares two measurements using the same units, rates compare two measurements with different units. Examples of rates would be $200 of work for 8 hours, or 500 miles per 20 gallons. Because the units are different, it is important to always include the units when discussing rates. Rates can be easily seen because if they are expressed in words, the two quantities are usually split up using one of the following words: *for, per, on, from, in.* Just as with ratios, it is important to write rates in lowest terms. A common rate that can be found in many real-life situations is cost per unit. This quantity describes how much one item or one unit costs. This rate allows the best buy to be determined, given a couple of different sizes of an item with different costs. For example, if 2 quarts of soup costs $3.50 and 3 quarts costs $4.60, to determine the best buy, the cost per quart should be found. $\frac{\$3.50}{2} = \$1.75$ per quart, and $\frac{\$4.60}{3} = \$1.53$ per quart. Therefore, the better deal would be the 3-quart option.

Rate of change problems involve calculating a quantity per some unit of measurement. Usually the unit of measurement is time. For example, meters per second is a common rate of change. To calculate this measurement, find the amount traveled in meters and divide by total time traveled. The calculation is an average of the speed over the entire time interval. Another common rate of change used in the real world is miles per hour. Consider the following problem that involves calculating an average rate of change in temperature. Last Saturday, the temperature at 1:00 a.m. was 34 degrees Fahrenheit, and at noon, the temperature had increased to 75 degrees Fahrenheit. What was the average rate of change over that time interval? The average rate of change is calculated by finding the change in temperature and dividing by the total hours elapsed. Therefore, the rate of change was equal to $\frac{75-34}{12-1} = \frac{41}{11}$ degrees per hour. This quantity rounded to two decimal places is equal to 3.72 degrees per hour.

A common rate of change that appears in algebra is the slope calculation. Given a linear equation in one variable, $y = mx + b$, the **slope**, m, is equal to $\frac{rise}{run}$ or $\frac{change\ in\ y}{change\ in\ x}$. In other words, slope is equivalent to the ratio of the vertical and horizontal changes between any two points on a line. The vertical change is known as the **rise,** and the horizontal change is known as the **run.** Given any two points on a line (x_1, y_1) and (x_2, y_2), slope can be calculated with the formula:

$$m = \frac{y_2 - y_1}{x_2 - x_1} = \frac{\Delta y}{\Delta x}$$

Common real-world applications of slope include determining how steep a staircase should be, calculating how steep a road is, and determining how to build a wheelchair ramp.

Many times, problems involving rates and ratios involve proportions. Recall that proportion states that two ratios (or rates) are equal. The property of cross products can be used to determine if a proportion is true, meaning both ratios are equivalent. If $\frac{a}{b} = \frac{c}{d}$, then to clear the fractions, multiply both sides by the least common denominator, bd. This results in $ac = cd$, which is equal to the result of multiplying along both diagonals. For example, $\frac{4}{40} = \frac{1}{10}$ yields the cross product $4 \cdot 10 = 40 \times 1$. $40 = 40$ shows that this proportion is true. Cross products are used when proportions are involved in real-world problems. Consider the following: If 3 pounds of fertilizer will cover 75 square feet of grass, how many pounds are needed for 375 square feet? To solve this problem, a proportion can be set up using two ratios. Let x equal the unknown quantity, pounds needed for 375 feet. Then, the equation found by setting the two given ratios equal to one another is $\frac{3}{75} = \frac{x}{375}$. Cross-multiplication gives $3 \times 375 = 75x$. Therefore, $1,125 = 75x$. Divide both sides by 75 to get $x = 15$. Therefore, 15 gallons of fertilizer are needed to cover 75 square feet of grass.

Another application of proportions involves similar triangles. If two triangles have the same measurement as two triangles in another triangle, the triangles are said to be **similar.** If two are the same, the third pair of angles are equal as well because the sum of all angles in a triangle is equal to 180 degrees. Each pair of equivalent angles are known as **corresponding angles. Corresponding sides** face the corresponding angles, and it is true that corresponding sides are in proportion. For example, consider the following set of similar triangles:

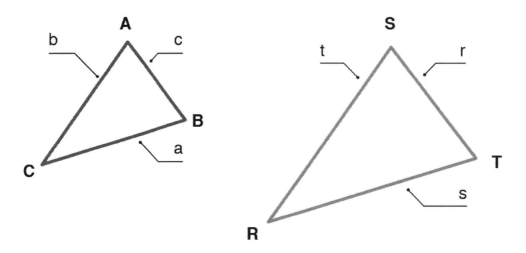

Angles A and S have the same measurement, angles C and R have the same measurement, and angles B and T have the same measurement. Therefore, the following proportion can be set up from the sides:

$$\frac{c}{r} = \frac{a}{s} = \frac{b}{t}$$

This proportion can be helpful in finding missing lengths in pairs of similar triangles. For example, if the following triangles are similar, a proportion can be used to find the missing side lengths, *a* and *b*.

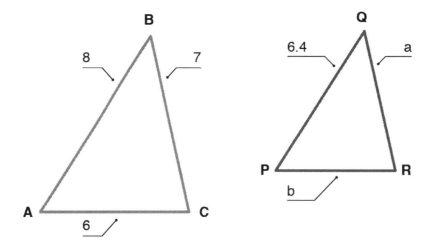

The proportions $\frac{8}{6.4} = \frac{6}{b}$ and $\frac{8}{6.4} = \frac{7}{a}$ can both be cross-multiplied and solved to obtain *a* = 5.6 and *b* = 4.8.

A real-life situation that uses similar triangles involves measuring shadows to find heights of unknown objects. Consider the following problem: A building casts a shadow that is 120 feet long, and at the same time, another building that is 80 feet high casts a shadow that is 60 feet long. How tall is the first building? Each building, together with the sun rays and shadows casted on the ground, forms a triangle. They are similar because each building forms a right angle with the ground, and the sun rays form equivalent angles. Therefore, these two pairs of angles are both equal. Because all angles in a triangle add up to 180 degrees, the third angles are equal as well. Both shadows form corresponding sides of the triangle, the buildings form corresponding sides, and the sun rays form corresponding sides. Therefore, the triangles are similar, and the following proportion can be used to find the missing building length:

$$\frac{120}{x} = \frac{60}{80}$$

Cross-multiply to obtain the cross products, $9600 = 60x$. Then, divide both sides by 60 to obtain $x = 160$. This solution means that the other building is 160 feet high.

Solving Unit Rate Problems

A **unit rate** is a rate with a denominator of one. It is a comparison of two values with different units where one value is equal to one. Examples of unit rates include 60 miles per hour and 200 words per minute. Problems involving unit rates may require some work to find the unit rate. For example, if Mary travels 360 miles in 5 hours, what is her speed, expressed as a unit rate? The rate can be expressed as the following fraction: $\frac{360\ miles}{5\ hours}$. The denominator can be changed to one by dividing by five. The numerator will also need to be divided by five to follow the rules of equality. This division turns the fraction into $\frac{72\ miles}{1\ hour}$, which can now be labeled as a unit rate because one unit has a value of one. Another type question involves the use of unit rates to solve problems. For example, if Trey needs to read 300 pages and his average speed is 75 pages per hour, will he be able to finish the reading in 5 hours? The unit rate is 75 pages per hour, so the total of 300 pages can be divided by 75 to find the time. After the division, the time it takes to read is four hours. The answer to the question is yes, Trey will finish the reading within 5 hours.

The Number System

Real Numbers

Whole numbers are the numbers 0, 1, 2, 3, ... Examples of other whole numbers would be 413 and 8,431. Notice that numbers such as 4.13 and $\frac{1}{4}$ are not included in whole numbers. **Counting numbers**, also known as **natural numbers**, consist of all whole numbers except for the zero. In set notation, the natural numbers are the set $\{1, 2, 3, ...\}$. The entire set of whole numbers and negative versions of those same numbers comprise the set of numbers known as **integers.** Therefore, in set notation, the integers are $\{..., -3, -2, -1, 0, 1, 2, 3, ...\}$. Examples of other integers are $-4,981$ and $90,131$. A number line is a great way to visualize the integers. Integers are labeled on the following number line:

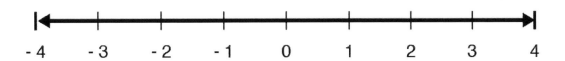

The arrows on the right- and left-hand sides of the number line show that the line continues indefinitely in both directions.

Fractions also exist on the number line as parts of a whole. For example, if an entire pie is cut into two pieces, each piece is half of the pie, or $\frac{1}{2}$. The top number in any fraction, known as the **numerator,** defines how many parts there are. The bottom number, known as the **denominator,** states how many pieces the whole is divided into. Fractions can also be negative or written in their corresponding decimal form. Examples of fractions are $\frac{2}{7}, \frac{-3}{14}$, and $\frac{14}{27}$.

A **decimal** is a number that uses a decimal point and numbers to the right of the decimal point representing the part of the number that is less than 1. For example, 3.5 is a decimal and is equivalent to the fraction $\frac{7}{2}$ or mixed number $3\frac{1}{2}$. The decimal is found by dividing 2 into 7.

Any number that can be expressed as a fraction is known as a **rational number.** Basically, if a and b are any integers and $b \neq 0$, then $\frac{a}{b}$ is a rational number. Rational numbers can be whole or negative numbers, fractions, or repeating decimals because these numbers can all be written as a fraction. Examples of rational numbers include $\frac{1}{2}, \frac{5}{4}$, and -8. Whole numbers can be written as fractions, where 25 and 17 can be written as $\frac{25}{1}$ and $\frac{17}{1}$. Any integer can be written as a fraction where the denominator is 1, so therefore the rational numbers consist of all fractions and all integers.

Any number that is not rational is known as an **irrational number.** Consider the number $\pi = 3.141592654 \ldots$. The decimal portion of that number extends indefinitely. In that situation, a number can never be written as a fraction. Another example of an irrational number is $\sqrt{2} = 1.414213662 \ldots$. Again, this number cannot be written as a ratio of two integers.

Together, the set of all rational and irrational numbers makes up the **real numbers.** The number line contains all real numbers. To graph a number other than an integer on a number line, it needs to be plotted between two integers. For example, 3.5 would be plotted halfway between 3 and 4.

Even numbers are integers that are divisible by 2. For example, 6, 100, 0, and -200 are all even numbers. **Odd numbers** are integers that are not divisible by 2. If an odd number is divided by 2, the result is a fraction. For example, -5, 11, and -121 are odd numbers.

Prime numbers consist of natural numbers greater than 1 that are not divisible by any other natural numbers other than themselves and 1. For example, 3, 5, and 7 are prime numbers. If a natural number is not prime, it is known as a **composite number.** 8 is a composite number because it is divisible by both 2 and 4, which are natural numbers other than itself and 1.

The **absolute value** of any real number is the distance from that number to 0 on the number line. The absolute value of a number can never be negative. For example, the absolute value of both 8 and -8 is 8 because they are both 8 units away from 0 on the number line. This is written as $|8| = |-8| = 8$.

Ordering and Comparing Rational Numbers
Ordering rational numbers is a way to compare two or more different numerical values. Determining whether two amounts are equal, less than, or greater than is the basis for comparing both positive and negative numbers. Also, a group of numbers can be compared by ordering them from the smallest amount to the largest amount. A few symbols are necessary to use when ordering rational numbers. The **equals sign**, =, shows that the two quantities on either side of the symbol have the same value. For example, $\frac{12}{3} = 4$ because both values are equivalent.

When the value of two numbers is not equivalent, an **inequality** exists. Inequalities use symbols to compare numbers as well. For example, < represents "less than." With this symbol, the smaller number is placed on the left and the larger number is placed on the right. Always remember that the symbol's "mouth" opens up to the larger number. When comparing negative and positive numbers, it is important to remember that the number occurring to the left on the number line is always smaller and is placed to the left of the symbol. This idea might seem confusing because some values could appear at first glance to be larger, even though they are not. For example, $-5 < 4$ is read "negative 5 is less than 4." Here is an image of a number line for help:

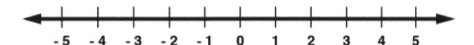

The symbol $\leq$ represents "less than or equal to," and it joins < with equality. Therefore, both $-5 \leq 4$ and $-5 \leq -5$ are true statements and "-5 is less than or equal to both 4 and -5." Other symbols are > and $\geq$, which represent "greater than" and "greater than or equal to." Both $4 \geq -1$ and $-1 \geq -1$ are correct ways to use these symbols.

Here is a chart of these four inequality symbols:

Symbol	Definition
<	less than
$\leq$	less than or equal to
>	greater than
$\geq$	greater than or equal to

Comparing integers is a straightforward process, especially when using the number line, but the comparison of decimals and fractions is not as obvious. When comparing two non-negative decimals, compare digit by digit, starting from the left. The larger value contains the first larger digit. For example, 0.1456 is larger than 0.1234 because the value 4 in the hundredths place in the first decimal is larger than the value 2 in the hundredths place in the second decimal. When comparing a fraction with a decimal, convert the fraction to a decimal and then compare in the same manner. Finally, there are a few options when comparing fractions. If two non-negative fractions have the same denominator, the fraction with the larger numerator is the larger value. If they have different denominators, they can be converted to equivalent fractions with a common denominator to be compared, or they can be converted to decimals to be compared. When comparing two negative decimals or fractions, a different approach must be used. It is important to remember that the smaller number exists to the left on the number line. Therefore, when comparing two negative decimals by place value, the number with the larger first place value is smaller due to the negative sign. Whichever value is closer to 0 is larger. For instance, -0.456 is larger than -0.498 because of the values in the hundredth places. If two negative fractions have the same denominator, the fraction with the larger numerator is smaller because of the negative sign.

Radicals and Exponents
Another operation that can be performed on rational numbers is the square root. Dealing with real numbers only, the **positive square root** of a number is equal to one of the two repeated positive factors of that number. For example:

$$\sqrt{49} = \sqrt{7 \times 7} = 7$$

A **perfect square** is a number that has a whole number as its square root. Examples of perfect squares are 1, 4, 9, 16, 25, etc. If a number is not a perfect square, an approximation can be used with a calculator. For example, $\sqrt{67} = 8.185$, rounded to the nearest thousandth place. The square root of a fraction involving perfect squares involves breaking up the problem into the square root of the numerator separate from the square root of the denominator.

For example:

$$\sqrt{\frac{16}{25}} = \frac{\sqrt{16}}{\sqrt{25}} = \frac{4}{5}$$

If the fraction does not contain perfect squares, a calculator can be used. Therefore, $\sqrt{\frac{2}{5}} = 0.632$, rounded to the nearest thousandth place. A common application of square roots involves the **Pythagorean theorem**. Given a right triangle, the sum of the squares of the two legs equals the square of the hypotenuse.

For example, consider the following right triangle:

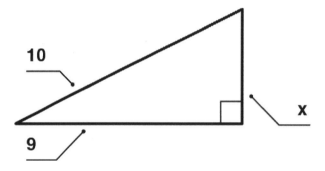

The missing side, x, can be found using the Pythagorean theorem.

$$9^2 + x^2 = 10^2$$

$$81 + x^2 = 100$$

$$x^2 = 19$$

To solve for x, take the square root of both sides. Therefore, $x = \sqrt{19} = 4.36$, which has been rounded to two decimal places.

In addition to the square root, the cube root is another operation. If a number is a **perfect cube**, the cube root of that number is equal to one of the three repeated factors. For example:

$$\sqrt[3]{27} = \sqrt[3]{3 \times 3 \times 3} = 3$$

Also, unlike square roots, a negative number has a cube root. The result is a negative number. For example:

$$\sqrt[3]{-27} = \sqrt[3]{(-3)(-3)(-3)} = -3$$

26

Similar to square roots, if the number is not a perfect cube, a calculator can be used to find an approximation. Therefore, $\sqrt[3]{\frac{2}{3}} = 0.873$, rounded to the nearest thousandth place.

Higher-order roots also exist. The number relating to the root is known as the **index.** Given the following root, $\sqrt[3]{64}$, 3 is the index, and 64 is the **radicand.** The entire expression is known as the **radical.** Higher-order roots exist when the index is larger than 3. They can be broken up into two groups: even and odd roots. Even roots, when the index is an even number, follow the properties of square roots. A negative number does not have an even root, and an even root is found by finding the single factor that is repeated the same number of times as the index in the radicand. For example, the fifth root of 32 is equal to 2 because:

$$\sqrt[5]{32} = \sqrt[5]{2 \times 2 \times 2 \times 2 \times 2} = 2$$

Odd roots, when the index is an odd number, follow the properties of cube roots. A negative number has an odd root. Similarly, an odd root is found by finding the single factor that is repeated that many times to obtain the radicand. For example, the 4th root of 81 is equal to 3 because $3^4 = 81$. This radical is written as $\sqrt[4]{81} = 4$.

The number 8 is rational because it can also be expressed as a fraction: $\frac{8}{1} = 8$. **Rational exponents** represent one way to show how roots are used to express multiplication of any number by itself. For example, 3^2 has a base of 3 and rational exponent of 2, or $\frac{2}{1}$. It can be rewritten as the square root of 3 raised to the first power, or $\sqrt[2]{3^1}$. Any number with a rational exponent can be written this way. The **numerator,** or number on top of the fraction, becomes the root and the **denominator,** or bottom number of the fraction, becomes the whole number exponent. Another example is $4^{\frac{3}{2}}$. It can be rewritten as the square root of four to the third power, or $\sqrt[2]{4^3}$. This can be simplified by performing the operations 4 to the third power, $4^3 = 4 \times 4 \times 4 = 64$, and then taking the square root of 64, $\sqrt[2]{64}$, which yields an answer of 8. Another way of stating the answer would be 4 to power $\frac{3}{2}$ is eight, or 4 times itself $\frac{3}{2}$ times is eight.

The nth root of a is given as $\sqrt[n]{a}$, which is called a **radical.** As just discussed, typical values for n are 2 and 3, which represent the square and cube roots. In this form, n represents an integer greater than or equal to 2, and a is a real number. If n is even, a must be nonnegative, and if n is odd, a can be any real number. This radical can be written in exponential form as $a^{\frac{1}{n}}$. Therefore, $\sqrt[4]{15}$ is the same as $15^{\frac{1}{4}}$ and $\sqrt[3]{-5}$ is the same as $(-5)^{\frac{1}{3}}$.

In a similar fashion, the nth root of a can be raised to a power m, which is written as $\left(\sqrt[n]{a}\right)^m$. This expression is the same as $\sqrt[n]{a^m}$. For example:

$$\sqrt[2]{4^3} = \sqrt[2]{64} = 8 = \left(\sqrt[2]{4}\right)^3 = 2^3$$

Because $\sqrt[n]{a} = a^{\frac{1}{n}}$, both sides can be raised to an exponent of m, resulting in:

$$\left(\sqrt[n]{a}\right)^m = \sqrt[n]{a^m} = a^{\frac{m}{n}}$$

This rule allows:

$$\sqrt[2]{4^3} = \left(\sqrt[2]{4}\right)^3 = 4^{\frac{3}{2}}$$

$$(2^2)^{\frac{3}{2}} = 2^{\frac{6}{2}} = 2^3 = 8$$

Negative exponents can also be incorporated into these rules. Any time an exponent is negative, the base expression must be flipped to the other side of the fraction bar and rewritten with a positive exponent. For instance:

$$2^{-3} = \frac{1}{2^3} = \frac{1}{8}$$

Therefore, two more relationships between radical and exponential expressions are:

$$a^{-\frac{1}{n}} = \frac{1}{\sqrt[n]{a}} \text{ and } a^{-\frac{m}{n}} = \frac{1}{\sqrt[n]{a^m}} = \frac{1}{\left(\sqrt[n]{a}\right)^m}$$

Thus:

$$8^{-3} = \frac{1}{\sqrt[3]{8}} = \frac{1}{2}$$

All of these relationships are very useful when simplifying complicated radical and exponential expressions. If an expression contains both forms, use one of these rules to change the expression to contain either all radicals or all exponential expressions. This process makes the entire expression much easier to work with, especially if the expressions are contained within equations.

Consider the following example:

$$\sqrt{x} \times \sqrt[4]{x}$$

It is written in radical form; however, it can be simplified into one radical by using exponential expressions first. The expression can be written as $x^{\frac{1}{2}} \times x^{\frac{1}{4}}$. It can be combined into one base by adding the exponents as:

$$x^{\frac{1}{2}+\frac{1}{4}} = x^{\frac{3}{4}}$$

Writing this back in radical form, the result is $\sqrt[4]{x^3}$.

Applying Estimation Strategies and Rounding Rules to Real-World Problems

Sometimes it is helpful to find an estimated answer to a problem rather than working out an exact answer. An estimation might be much quicker to find, and given the scenario, an estimation might be all that is required. For example, if Aria goes grocery shopping and has only a $100 bill to cover all of her purchases, it might be appropriate for her to estimate the total of the items she is purchasing to determine if she has enough money to cover them. Also, an estimation can help determine if an answer makes sense. For instance, if an answer in the 100s is expected, but the result is a fraction less than 1, something is probably wrong in the calculation.

The first type of estimation involves rounding. As mentioned, **rounding** consists of expressing a number in terms of the nearest decimal place like the tenth, hundredth, or thousandth place, or in terms of the nearest whole number unit like tens, hundreds, or thousands place. When rounding to a specific place value, look at the digit to the right of the place. If it is 5 or higher, round the number to its left up to the next value, and if it is 4 or lower, keep that number at the same value. For instance, 1,654.2674 rounded to the nearest thousand is 2,000, and the same number rounded to the nearest thousandth is 1,654.267. Rounding can be used in the scenario when grocery totals need to be estimated. Items can be rounded to the nearest dollar. For example, a can of corn that costs $0.79 can be rounded to $1.00, and then all other items can be rounded in a similar manner and added together. When working with larger numbers, it might make more sense to round to higher place values. For example, when estimating the total value of a dealership's car inventory, it would make sense to round the car values to the nearest thousands place.

The price of a car that is on sale for $15,654 can be estimated at $16,000. All other cars on the lot could be rounded in the same manner, and then their sum can be found. Depending on the situation, it might make sense to calculate an over-estimate. For example, to make sure Aria has enough money at the grocery store, rounding up every time for each item would ensure that she will have enough money when it comes time to pay. A $0.40 item rounded up to $1.00 would ensure that there is a dollar to cover that item. Traditional rounding rules would round $0.40 to $0, which does not make sense in this particular real-world setting. Aria might not have a dollar available at checkout to pay for that item if she uses traditional rounding. It is up to the customer to decide the best approach when estimating.

Rounding is useful in both approximating an answer when an exact answer is not needed and for comparison when an exact answer is needed. For instance, if you had a complicated set of operations to complete and your estimate was $1,000, if you obtained an exact answer of $100,000, something is off. You might want to check your work to see if a mistake was made because an estimate should not be that different from an exact answer. Estimates can also be helpful with square roots. If a square root of a number is not known, the closest perfect square can be found for an approximation. For example, $\sqrt{50}$ is not equal to a whole number, but 50 is close to 49, which is a perfect square, and $\sqrt{49} = 7$. Therefore, $\sqrt{50}$ is a little bit larger than 7. The actual approximation, rounded to the nearest thousandth, is 7.071.

Estimating is also very helpful when working with measurements. Bryan is updating his kitchen and wants to retile the floor. Again, an over-measurement might be useful. Also, rounding to nearest half-unit might be helpful. For instance, one side of the kitchen might have an exact measurement of 14.32 feet, and the most useful measurement needed to buy tile could be estimating this quantity to be 14.5 feet. If the kitchen was rectangular and the other side measured 10.9 feet, Bryan might round the other side to 11 feet. Therefore, Bryan would find the total tile necessary according to the following area calculation: $14.5 \times 11 = 159.5$ square feet. To make sure he purchases enough tile, Bryan would probably want to purchase at least 160 square feet of tile. This is a scenario in which an estimation might be more useful than an exact calculation. Having more tile than necessary is better than having an exact amount, in case any tiles are broken or otherwise unusable.

Finally, estimation is helpful when exact answers are necessary. Consider a situation in which Sabina has many operations to perform on numbers with decimals, and she is allowed a calculator to find the result. Even though an exact result can be obtained with a calculator, there is always a possibility that Sabina could make an error while inputting the data. For example, she could miss a decimal place, or misuse a parenthesis, causing a problem with the actual order of operations. In this case, a quick estimation at the beginning would be helpful to make sure the final answer is given with the correct number of units. Sabina has to find the exact total of 10 cars listed for sale at the dealership. Each price has two decimal places included to account for both dollars and cents. If one car is listed at $21, 234.43 but Sabina incorrectly inputs into the calculator the price of $2,123.443, this error would throw off the final sum by almost $20,000. A quick estimation at the beginning, by rounding each price to the nearest thousands place and finding the sum of the prices, would give Sabina an amount to compare the exact amount to. This comparison would let Sabina see if an error was made in her exact calculation.

Algebra

Algebraic Expressions

Applying an Understanding of Arithmetic to Algebraic Expressions
Interpreting Parts of an Expression
An **algebraic expression** is a mathematical phrase that may contain numbers, variables, and mathematical operations. An expression represents a single quantity. For example, $3x + 2$ is an algebraic expression.

An **algebraic equation** is a mathematical sentence with two expressions that are equal to each other. That is, an equation must contain an *equals* sign, as in $3x + 2 = 17$. This statement says that the value of the expression on the left side of the equals sign is equivalent to the value of the expression on the right side. In an expression, there are not two sides because there is no equals sign. The equals sign ($=$) is the difference between an expression and an equation.

To distinguish an expression from an equation, just look for the *equals sign*.

Example: Determine whether each of these is an expression or an equation.

1. $16 + 4x = 9x - 7$ Solution: Equation

2. $-27x - 42 + 19y$ Solution: Expression

3. $4 = x + 3$ Solution: Equation

Adding and Subtracting Linear Algebraic Expressions
To add and subtract linear algebra expressions, you must combine like terms. **Like terms** are described as those terms that have the same variable with the same exponent. In the following example, the x-terms can be added because the variable is the same and the exponent on the variable of one is also the same. These terms add to be $9x$. The other like terms are called **constants** because they have no variable component. These terms will add to be nine.

Example: Add $(3x - 5) + (6x + 14)$

                         $3x - 5 + 6x + 14$ Rewrite without parentheses

                         $3x + 6x - 5 + 14$ Commutative Property of Addition

                         $9x + 9$ Combine like terms

When subtracting linear expressions, be careful to add the opposite when combining like terms. Do this by distributing -1, which is multiplying each term inside the second parenthesis by negative one. Remember that distributing -1 changes the sign of each term.

Example: Subtract $(17x + 3) - (27x - 8)$

$17x + 3 - 27x + 8$	Distributive Property
$17x - 27x + 3 + 8$	Commutative Property of Addition
$-10x + 11$	Combine like terms

Example: Simplify by adding or subtracting: $(6m + 28z - 9) + (14m + 13) - (-4z + 8m + 12)$

$6m + 28z - 9 + 14m + 13 + 4z - 8m - 12$	Distributive Property
$6m + 14m - 8m + 28z + 4z - 9 + 13 - 12$	Commutative Property of Addition
$12m + 32z - 8$	Combine like terms

Evaluating and Simplifying Algebraic Expressions
When asked to evaluate for given values of variable(s), a numerical value with be the answer. For example:

$$\text{Evaluate: } a - 2b + ab \text{ for } a = 3 \text{ and } b = -1$$

To evaluate an expression, the given values should be substituted for the variables and simplified using the order of operations. In this case:

$$(3) - 2(-1) + (3)(-1)$$

Parentheses are used when substituting.

Simplifying algebraic expressions requires combining like terms. A **term** is a number, variable, or product of a number and variables separated by addition and subtraction. Consider the following: Simplify:

$$5x^2 - 10x + 2 - 8x^2 + x - 1$$

The terms in the above expressions are: $5x^2, -10x, 2, -8x^2, x$, and -1. **Like terms** have the same variables raised to the same powers (exponents). To combine like terms, the coefficients (numerical factor of the term including sign) are added, while the variables and their powers are kept the same. The example above simplifies to:

$$-3x^2 - 9x + 1$$

Solving Problems Using Numerical and Algebraic Expressions
Translating Phrases and Sentences into Expressions, Equations, and Inequalities
When presented with a real-world problem that must be solved, the first step is always to determine what the unknown quantity is that must be solved for. Use a **variable**, such as x or t, to represent that unknown quantity. Sometimes there can be two or more unknown quantities. In this case, either choose an additional variable, or if a relationship exists between the unknown quantities, express the other quantities in terms of the original variable. After choosing the variables, form algebraic expressions and/or equations

31

that represent the verbal statement in the problem. The following table shows examples of vocabulary used to represent the different operations:

Addition	Sum, plus, total, increase, more than, combined, in all
Subtraction	Difference, less than, subtract, reduce, decrease, fewer, remain
Multiplication	Product, multiply, times, part of, twice, triple
Division	Quotient, divide, split, each, equal parts, per, average, shared

Most of the time, these words and phrases are combined to represent expressions that deal with one or more operation. For example, "ten subtracted from nine times a number" would be represented as $9x - 10$, and "the quotient of a number and 7 increased by 8" would be represented as $\frac{x}{7} + 8$. The word problems that typically use these expressions will have a statement of equality. For instance, the problem could say "ten subtracted from nine times a number equals 20; find that number." In this case, the algebraic expression shown previously would be set equal to 20 and then solved for x.

$$9x - 10 = 20$$

Add ten to both sides and then divide by 9 to get the solution $x = \frac{30}{9}$, which reduces to $x = \frac{10}{3}$.

Essentially, translating sentences describing relationships between variables and constants to algebraic expressions and equations involves recognizing key words that represent mathematical operations. This process is known as **modeling.** The combination of operations and variables form both mathematical expression and equations. The difference between expressions and equations are that there is no equals sign in an expression, and that expressions are **evaluated** to find an unknown quantity, while equations are **solved** to find an unknown quantity. Also, inequalities can exist within verbal mathematical statements. Instead of a statement of equality, expressions state quantities are *less than, less than or equal to, greater than,* or *greater than or equal to.* Another type of inequality is when a quantity is said to be *not equal to* another quantity. The symbol used to represent "not equal to" is ≠.

The steps for solving inequalities in one variable are the same steps for solving equations in one variable. The addition and multiplication principles are used. However, to maintain a true statement when using the $<, \leq, >$, and $\geq$ symbols, if a negative number is either multiplied times both sides of an inequality or divided from both sides of an inequality, the sign must be flipped. For instance, consider the following inequality: $3 - 5x \leq 8$. First, 3 is subtracted from each side to obtain $-5x \leq 5$. Then, both sides are divided by -5, while flipping the sign, to obtain $x \geq -1$. Therefore, any real number greater than or equal to -1 satisfies the original inequality.

Other types of expressions, besides linear expressions, can be the results of modeling. If the variable is raised to a power other than 1, the result is a **polynomial expression**. The path of an object thrown up into the air is a common example of this. The graph of an object represents an upside-down parabola, which is modeled by an equation of the type $y = -ax^2$. In this case, a represents the height of the object at its highest point before coming back down to the ground, and the negative sign shows that the parabola is upside down. Here is the graph of a parabola:

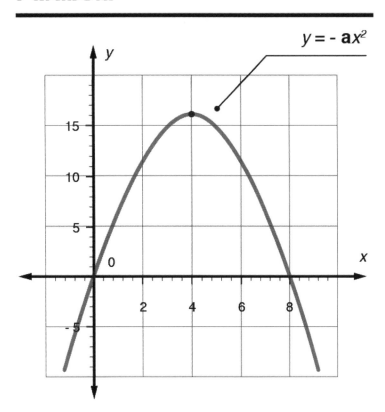

Parabola

$y = -ax^2$

Rewriting Simple Rational Expressions
A **rational expression** is a fraction or a ratio in which both the numerator and denominator are polynomials that are not equal to zero. A **polynomial** is a mathematical expression containing the sum and difference of one or more terms that are constants multiplied times variables raised to positive powers. Here are some examples of rational expressions:

$$\frac{2x^2+6x}{x}, \quad \frac{x-2}{x^2-6x+8}, \quad \text{and} \quad \frac{x+2}{x^3-1}$$

Such expressions can be simplified using different forms of division. The first example can be simplified in two ways. First, because the denominator is a monomial, the expression can be split up into two expressions: $\frac{2x^2}{x} + \frac{6x}{x}$, and then simplified using properties of exponents as $2x + 6$. It also can be simplified using factoring and then crossing common factors out of the numerator and denominator. For instance, it can be written as:

$$\frac{2x(x+3)}{x} = 2(x+3) = 2x + 6$$

The second expression above can also be simplified using factoring. It can be written as:

$$\frac{x-2}{(x-2)(x-4)} = \frac{1}{x-4}$$

Finally, the third example can only be simplified using long division, as there are no common factors in the numerator and denominator. First, divide the first term of the denominator by the first term of the numerator, then write that in the quotient. Then, multiply the divisor by that number and write it below the dividend. Subtract, bring down the next term from the dividend, and continue that process with the next first term and first term of the divisor. Continue the process until every term in the divisor is accounted for. Here is the actual long division:

Simplifying Expressions Using Long Division

$$
\require{enclose}
\begin{array}{rll}
& x^2 \quad -2x \quad +4 \\
x+2 \enclose{longdiv}{x^3 \qquad\qquad\qquad -1} \\
\underline{x^3 \;+2x^2} \\
-2x^2 \qquad\qquad -1 \\
\underline{-2x^2 \;-4x} \\
4x \quad -1 \\
\underline{4x \;+8} \\
-9
\end{array}
$$

Interpreting Parts of Nonlinear Expressions in Terms of Their Context
If a quantity increases or decreases at a constant rate as another quantity increases, then this idea is represented as a **linear expression**, and the graph of such a relationship is a straight line. All other relationships are nonlinear, and nonlinear expressions must be used as mathematical representations of such instances.

Common nonlinear relationships that exist between two quantities are **inverse variation equations**, which are represented by equations such as $y = \frac{k}{x}$ and $y = \frac{k}{x^2}$ with constants k, **quadratic equations** of the form $y = ax^2 + bx + c$, and **exponential equations** of the form $y = a^x$.

Inverse variation situations arise when, as one quantity increases, the other quantity decreases in proportion. For instance, as a person increases the speed of a car she is driving, the time it takes to reach the destination decreases. This is a nonlinear relationship regarding inverse variation.

Recall that quadratic equations are used to model something shaped like a parabola. For instance, if a ball is thrown into the air, it travels higher and higher and eventually slows down to reach its highest point, then stops dropping at a faster and faster rate. A quadratic equation must be used to tell the position of the ball given the amount of time since the ball was thrown. This relationship is nonlinear.

Finally, an exponential equation is used to model something with exponential growth or decay. If something grows exponentially, such as compound interest, the amount is multiplied times a growth factor for every increase in x. If something decays exponentially, the amount is multiplied times a factor between 0 and 1 for every increase in x. When a population is declining, an exponential decay equation can be used to represent the situation.

Generating Equivalent Expressions
Creating Equivalent Expressions Involving Rational Exponents and Radicals
Writing radical expressions into equivalent forms involving rational exponents can help in simplifying complex radical expressions. The rule that helps this conversion is:

$$\sqrt[n]{x^m} = x^{\frac{m}{n}}$$

If $m = 1$, the rule is simply $\sqrt[n]{x} = x^{\frac{1}{n}}$. For instance, consider the following expression:

$$\sqrt[4]{x}\sqrt[2]{y}$$

It can be written as one radical expression, but first it needs to be converted to an equivalent expression using rational expressions. The equivalent expression is $x^{\frac{1}{4}}y^{\frac{1}{2}}$. The goal is to have one radical, which means one index n, so a common denominator of the exponents must be found. The common denominator is 4, so an equivalent expression is $x^{\frac{1}{4}}y^{\frac{2}{4}}$. The exponential rule $a^m b^m = (ab)^m$ can be used to, in a sense, factor out a $\frac{1}{4}$ out of both exponents. This process results in the expression $(xy^2)^{\frac{1}{4}}$, and its equivalent radical form is $\sqrt[4]{xy^2}$. Converting to rational exponents has allowed the entire expression to be written as one radical.

Another type of problem could involve going in the opposite direction—starting with rational exponents and using an equivalent radical form to simplify the expression. For instance, $32^{\frac{1}{5}}$ might not seem obviously equal to 2. However, putting it in is equivalent radical form $\sqrt[5]{32}$ shows that it is equivalent to the fifth root of 32, which is 2.

Creating an Equivalent Form of an Algebraic Expression
Two algebraic expressions are equivalent if, even though they look different, they represent the same expression. Therefore, plugging in the same values into the variables in each expression will result in the same result in both expressions. To obtain an equivalent form of an algebraic expression, laws of algebra must be followed. For instance, addition and multiplication are both commutative and associative. Therefore, terms in an algebraic expression can be added in any order and multiplied in any order. For instance, $4x + 2y$ is equivalent to $2y + 4x$ and $y \times 2 + x \times 4$. Also, the distributive property allows a number to be distributed throughout parentheses, as in the following:

$$a(b + c) = ab + ac$$

The two expressions on both sides of the equals sign are equivalent. Also, collecting like terms is important when working with equivalent forms. The simplest version of an expression is always the one easiest to work with, so all like terms (those with the same variables raised to the same powers) must be combined.

Note that an expression is not an equation; therefore, expressions cannot be multiplied by numbers, divided by numbers, or have numbers added to them or subtracted from them and still have equivalent expressions. These processes can only happen in equations when the same step is performed on both sides of the equals sign.

Using the Distributive Property to Generate Equivalent Linear Algebraic Expressions
The **distributive property** is a way of taking a factor and multiplying it through a given expression in parentheses. Each term inside the parentheses is multiplied by the outside factor, eliminating the parentheses. The following example shows how to distribute the number 3 to all the terms inside the parentheses. The general principle of the distributive property is modeled by the following formula: $a(b + c) = ab + ac$.

Example: Use the distributive property to write an equivalent algebraic expression:

$$3(2x + 7y + 6)$$

$$3(2x) + 3(7y) + 3(6) \qquad \text{Distributive property}$$

$$6x + 21y + 18 \qquad \text{Simplify}$$

Because $a - b$ can be written $a + (-b)$, the distributive property can be applied in the example below:

Example: Use the distributive property to write an equivalent algebraic expression.

$$7(5m - 8)$$

$$7[5m + (-8)] \qquad \text{Rewrite subtraction as addition of } -8$$

$$7(5m) + 7(-8) \qquad \text{Distributive property}$$

$$35m - 56 \qquad \text{Simplify}$$

In the following example, note that the factor of 2 is written to the right of the parentheses but is still distributed as before.

Example: Use the distributive property to write an equivalent algebraic expression:

$$(3m + 4x - 10)2$$

$$(3m)2 + (4x)2 + (-10)2 \qquad \text{Distributive property}$$

$$6m + 8x - 20 \qquad \text{Simplify}$$

Example: $-(-2m + 6x)$

In this example, the negative sign in front of the parentheses can be interpreted as $-1(-2m + 6x)$

$-1(-2m + 6x)$

$-1(-2m) + (-1)(6x)$ Distributive property

$2m - 6x$ Simplify

Determining the Most Suitable Form of an Expression or Equation
When given a problem, it is necessary to determine the best form of an expression or equation to use, given the context. Usually this involves some algebraic manipulation. If an equation is given, the simplest form of the equation is best. Simplifying involves using the distributive property, collecting like terms, etc., If an equation is needed to be solved, properties involving performing the same operation on both sides of the equation must be used. For instance, if a number is added to one side of the equals sign, it must be added to the other side as well. This maintains a true equation.

If an expression is given, simplifying can only involve properties allowing to rewrite the expression as an equivalent form. If there is no equals sign, mathematical operations cannot be performed on the expression, unless it is a rational expression. A rational expression can be written in the form of a fraction, in which the numerator and denominator are both polynomials and the denominator is not equal to zero. Rational expressions can always be multiplied times a form of 1. For example, consider the following rational expression involving radicals: $\frac{2}{\sqrt{2}}$. It is incorrect to write a fraction with a root in the denominator, and therefore the expression must be rationalized. Multiply the fraction times $\frac{\sqrt{2}}{\sqrt{2}}$, a form of 1. This results in $\frac{2}{\sqrt{2}} \times \frac{\sqrt{2}}{\sqrt{2}} = \frac{2\sqrt{2}}{\sqrt{4}} = \frac{2\sqrt{2}}{2} = \sqrt{2}$, which is the most suitable form of the expression.

Algebraic Equations and Inequalities

The Connection Between Proportional Relationships and Linear Equations
Linear growth involves a quantity—the **dependent variable**—increasing or decreasing at a constant rate as another quantity—the **independent variable**—increases as well. The graph of linear growth is a straight line. Linear growth is represented as the following equation: $y = mx + b$, where m is the **slope** of the line, also known as the **rate of change**, and b is the **y-intercept**. If the y-intercept is 0, then the linear growth is actually known as **direct variation**. If the slope is positive, the dependent variable increases as the independent variable increases, and if the slope is negative, the dependent variable decreases as the independent variable increases.

A linear relationship between two quantities takes the standard equation form of $y = mx + b$, or in function form, $f(x) = mx + b$. In a linear function, the value of y depends on the value of x, and y increases or decreases at a constant rate as x increases. Therefore, the independent variable is x, and the dependent variable is y. Note that a function is a relationship or equation where every x-value (called the **domain**) has exactly one y-value (called the **range**). The graph of a linear function is a line, and the constant rate can be seen by looking at the steepness, or slope, of the line. If the line increases from left to right, the slope is positive. If the line slopes downward from left to right, the slope is negative. In the function, m represents slope. Each point on the line is an **ordered pair** (x, y), where x represents the x-coordinate of the point and y represents the y-coordinate of the point. The point where $x = 0$ is known as the y-intercept, and it is the place where the line crosses the y-axis. If $x = 0$ is plugged into $f(x) = mx + b$, the result is $f(0) = b$, so therefore, the point $(0, b)$ is the y-intercept of the line. The derivative of a linear function is its slope.

Consider the following situation. A taxicab driver charges a flat fee of $2 per ride and $3 a mile. This statement can be modeled by the function $f(x) = 3x + 2$ where x represents the number of miles and $f(x) = y$ represents the total cost of the ride. The total cost increases at a constant rate of $2 per mile, and that is why this situation is a linear relationship. The slope $m = 3$ is equivalent to this rate of change. The flat fee of $2 is the y-intercept. It is the place where the graph crosses the x-axis, and it represents the cost when $x = 0$, or when no miles have been traveled in the cab. The y-intercept in this situation represents the flat fee.

One-Variable Linear Equations and Inequalities
Solving Equations in One Variable

An **equation in one variable** is a mathematical statement where two algebraic expressions in one variable, usually x, are set equal. To solve the equation, the variable must be isolated on one side of the equals sign. The addition and multiplication principles of equality are used to isolate the variable. The **addition principle of equality** states that the same number can be added to or subtracted from both sides of an equation. Because the same value is being used on both sides of the equals sign, equality is maintained. For example, the equation $2x - 3 = 5x$ is equivalent to both $2x - 3 + 2 = 5x + 2$, and $2x - 3 - 5 = 5x - 5$. This principle can be used to solve the following equation: $x + 5 = 4$. The variable x must be isolated, so to move the 5 from the left side, subtract 5 from both sides of the equals sign. Therefore:

$$x + 5 - 5 = 4 - 5$$

So, the solution is $x = -1$.

This process illustrates the idea of an **additive inverse** because subtracting 5 is the same as adding -5. Basically, add the opposite of the number that must be removed to both sides of the equals sign. The **multiplication principle of equality** states that equality is maintained when a number is either multiplied times both expressions on each side of the equals sign, or when both expressions are divided by the same number. For example, $4x = 5$ is equivalent to both $16x = 20$ and $x = \frac{5}{4}$. Multiplying both sides times 4 and dividing both sides by 4 maintains equality. Solving the equation $6x - 18 = 5$ requires the use of both principles. First, apply the addition principle to add 18 to both sides of the equals sign, which results in $6x = 23$. Then use the multiplication principle to divide both sides by 6, giving the solution $x = \frac{23}{6}$. Using the multiplication principle in the solving process is the same as involving a multiplicative inverse. A **multiplicative inverse** is a value that, when multiplied by a given number, results in 1. Dividing by 6 is the same as multiplying by $\frac{1}{6}$, which is both the reciprocal and multiplicative inverse of 6.

When solving a linear equation in one variable, checking the answer shows if the solution process was performed correctly. Plug the solution into the variable in the original equation. If the result is a false statement, something was done incorrectly during the solution procedure. Checking the example above gives the following:

$$6 \times \frac{23}{6} - 18 = 23 - 18 = 5$$

Therefore, the solution is correct.

Some equations in one variable involve fractions or the use of the distributive property. In either case, the goal is to obtain only one variable term and then use the addition and multiplication principles to isolate that variable. Consider the equation $\frac{2}{3}x = 6$. To solve for x, multiply each side of the equation by the reciprocal of $\frac{2}{3}$, which is $\frac{3}{2}$. This step results in $\frac{3}{2} \times \frac{2}{3}x = \frac{3}{2} \times 6$, which simplifies into the solution $x = 9$. Now consider the equation:

$$3(x + 2) - 5x = 4x + 1$$

Use the distributive property to clear the parentheses. Therefore, multiply each term inside the parentheses by 3. This step results in:

$$3x + 6 - 5x = 4x + 1$$

Next, collect like terms on the left-hand side. Remember that **like terms** are terms with the same variable or variables raised to the same exponent(s). Only like terms can be combined through addition or subtraction. After collecting like terms, the equation is:

$$-2x + 6 = 4x + 1$$

Finally, apply the addition and multiplication principles. Add $2x$ to both sides to obtain $6 = 6x + 1$. Then, subtract 1 from both sides to obtain $5 = 6x$. Finally, divide both sides by 6 to obtain the solution $\frac{5}{6} = x$.

Two other types of solutions can be obtained when solving an equation in one variable. The final result could be that there is either no solution or that the solution set contains all real numbers. Consider the equation:

$$4x = 6x + 5 - 2x$$

First, the like terms can be combined on the right to obtain $4x = 4x + 5$. Next, subtract $4x$ from both sides. This step results in the false statement $0 = 5$. There is no value that can be plugged into x that will ever make this equation true. Therefore, there is no solution. The solution procedure contained correct steps, but the result of a false statement means that no value satisfies the equation. The symbolic way to denote that no solution exists is ∅.

Next, consider the equation:

$$5x + 4 + 2x = 9 + 7x - 5$$

Combining the like terms on both sides results in:

$$7x + 4 = 7x + 4$$

The left-hand side is exactly the same as the right-hand side. Using the addition principle to move terms, the result is $0 = 0$, which is always true. Therefore, the original equation is true for any number, and the solution set is all real numbers. The symbolic way to denote such a solution set is $\mathbb{R}$, or in interval notation, $(-\infty, \infty)$.

Solving a Linear Inequality in One Variable
A **linear equation** in x can be written in the form $ax + b = 0$. A **linear inequality** is very similar, although the equals sign is replaced by an inequality symbol such as $<, >, \leq,$ or $\geq$. In any case, a can never be 0. Some examples of linear inequalities in one variable are $2x + 3 < 0$ and $4x - 2 \leq 0$. Solving an inequality involves finding the set of numbers that when plugged into the variable, make the inequality a true statement. These numbers are known as the **solution set** of the inequality.

To solve an inequality, use the same properties that are necessary in solving equations. First, add or subtract variable terms and/or constants to obtain all variable terms on one side of the equals sign and all constant terms on the other side. Then, either multiply both sides times the same number, or divide both sides by the same number, to obtain an inequality that gives the solution set. When multiplying times, or dividing by, a negative number in an inequality, change the direction of the inequality symbol. The solution set can be graphed on a number line. Consider the linear inequality $-2x - 5 > x + 6$. First, add 5 to both sides and subtract $-x$ off of both sides to obtain $-3x > 11$. Then, divide both sides by -3, making sure to change the direction of the inequality symbol. These steps result in the solution $x < -\frac{11}{3}$. Therefore, any number less than $-\frac{11}{3}$ satisfies this inequality.

Algebraically Solving Linear Equations or Inequalities in One Variable
A linear equation in one variable can be solved using the following steps:

> 1. Simplify the algebraic expressions on both sides of the equals sign by removing all parentheses, using the distributive property, and then collect all like terms.

> 2. Collect all variable terms on one side of the equals sign and all constant terms on the other side by adding the same quantity to both sides of the equals sign, or by subtracting the same quantity from both sides of the equals sign.

> 3. Isolate the variable by either dividing both sides of the equation by the same number, or by multiplying both sides by the same number.

> 4. Check the answer.

The only difference between solving linear inequalities versus equations is that when multiplying by a negative number or dividing by a negative number, the direction of the inequality symbol must be reversed.

If an equation contains multiple fractions, it might make sense to clear the equation of fractions first by multiplying all terms by the least common denominator. Also, if an equation contains several decimals, it might make sense to clear the decimals as well by multiplying times a factor of 10. If the equation has decimals in the hundredth place, multiply every term in the equation by 100.

<u>Systems of Equations and Inequalities</u>
Systems of Two Linear Equations in Two Variables
An example of a system of two linear equations in two variables is the following:

$$2x + 5y = 8$$

$$5x + 48y = 9$$

A solution to a system of two linear equations is an ordered pair that satisfies both the equations in the system. A system can have one solution, no solution, or infinitely many solutions. The solution can be found through a graphing technique. The solution of a system of equations is actually equal to the point of intersection of both lines. If the lines intersect at one point, there is one solution and the system is said to be **consistent**. However, if the two lines are parallel, they will never intersect and there is no solution. In this case, the system is said to be **inconsistent.** Third, if the two lines are actually the same line, there are infinitely many solutions and the solution set is equal to the entire line. The lines are dependent. Here is a summary of the three cases:

Consistent	Inconsistent	Dependent
One solution	No solution	Infinite number of solutions
Lines intersect	Lines are parallel	Coincide/Same line

Consider the following system of equations:

$$y + x = 3$$

$$y - x = 1$$

To find the solution graphically, graph both lines on the same xy-plane. Graph each line using either a table of ordered pairs, the x- and y-intercepts, or slope and the y-intercept. Then, locate the point of intersection.

The graph is shown here:

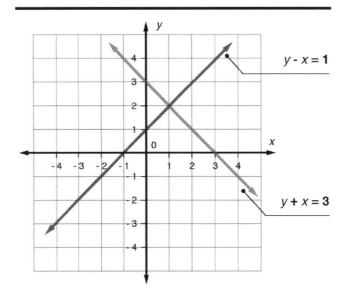

The System of Equations $\begin{cases} y + x = 3 \\ y - x = 1 \end{cases}$

It can be seen that the point of intersection is the ordered pair (1, 2). This solution can be checked by plugging it back into both original equations to make sure it results in true statements. This process results in:

$$2 + 1 = 3$$

$$2 - 1 = 1$$

Both are true equations, so therefore the point of intersection is truly the solution.

The following system has no solution:

$$y = 4x + 1$$

$$y = 4x - 1$$

Both lines have the same slope and different *y*-intercepts, and therefore they are parallel. This means that they run alongside each other and never intersect.

Finally, the following solution has infinitely many solutions:

$$2x - 7y = 12$$

$$4x - 14y = 24$$

Note that the second equation is equal to the first equation times 2. Therefore, they are the same line. The solution set can be written in set notation as $\{(x, y)|2x - 7y = 12\}$, which represents the entire line.

Algebraically Solving Systems of Two Linear Equations in Two Variables
There are two algebraic methods to finding solutions. The first is **substitution**. This process is better suited for systems when one of the equations is already solved for one variable, or when solving for one variable is easy to do. The equation that is already solved for is substituted into the other equation for that variable, and this process results in a linear equation in one variable. This equation can be solved for the given variable, and then that solution can be plugged into one of the original equations, which can then be solved for the other variable. This last step is known as **back-substitution** and the end result is an ordered pair.

A system that is best suited for substitution is the following:

$$y = 4x + 2$$

$$2x + 3y = 9$$

The other method is known as **elimination,** or the **addition method**. This is better suited when the equations are in standard form $Ax + By = C$. The goal in this method is to multiply one or both equations times numbers that result in opposite coefficients. Then, add the equations together to obtain an equation in one variable. Solve for the given variable, then take that value and back-substitute to obtain the other part of the ordered pair solution.

A system that is best suited for elimination is the following:

$$2x + 3y = 8$$

$$4x - 2y = 10$$

Note that in order to check an answer when solving a system of equations, the solution must be checked in both original equations to show that it solves not only one of the equations, but both of them.

If throughout either solution procedure the process results in an untrue statement, there is no solution to the system. Finally, if throughout either solution procedure the process results in the variables dropping out, which gives a statement that is always true, there are infinitely many solutions.

Systems of Linear Inequalities in Two Variables
A system of linear inequalities in two variables consists of two inequalities in two variables, x and y. For example, the following is a system of linear inequalities in two variables:

$$\begin{cases} 4x + 2y < 1 \\ 2x - y \le 0 \end{cases}$$

The curly brace on the left side shows that the two inequalities are grouped together. A solution of a single inequality in two variables is an ordered pair that satisfies the inequality. For example, (1, 3) is a solution of the linear inequality $y \ge x + 1$ because when plugged in, it results in a true statement. The graph of an inequality in two variables consists of all ordered pairs that make the solution true. Therefore, the entire solution set of a single inequality contains many ordered pairs, and the set can be graphed by using a half plane. A **half plane** consists of the set of all points on one side of a line. If the inequality consists of > or <, the line is dashed because no solutions actually exist on the line shown. If the inequality consists of $\ge$ or $\le$, the line is solid and solutions are on the line shown. To graph a linear inequality, graph the corresponding equation found by replacing the inequality symbol with an equals sign. Then pick a test point that exists on either side of the line. If that point results in a true statement when plugged into the original inequality, shade in the side containing the test point. If it results in a false statement, shade in the opposite side.

Solving Systems of Linear Inequalities Graphically
Solving a system of linear inequalities must be done graphically. Follow the process as described above for both given inequalities. The solution set to the entire system is the region that is in common to every graph in the system. For example, here is the solution to the following system:

$$\begin{cases} y \geq 3 - x \\ y \leq -3 - x \end{cases}$$

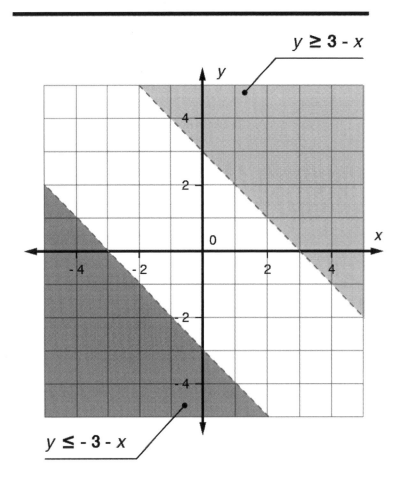

The solution to $\begin{cases} y \geq 3 - x \\ y \leq -3 - x \end{cases}$

Note that there is no region in common, so this system has no solution.

Number and Shape Patterns

Patterns in math are those sets of numbers or shapes that follow a rule. Given a set of values, patterns allow the question of "what's next?" to be answered. In the following set there are two types of shapes, a lighter rectangle and a darker circle. The set contains a pattern because every odd-placed shape is a lighter rectangle and every even-placed spot is taken by a darker circle. This is a pattern because there is a rule of lighter rectangle, then darker circle, that is followed to find the set.

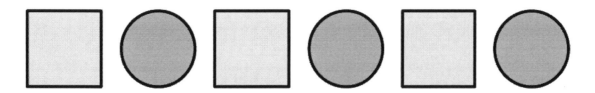

A set of numbers can also be described as having a pattern if there is a rule that can be followed to reproduce the set. The following set of numbers has a rule of adding 3 each time. It begins with zero and increases by 3 each time. By following this rule and pattern, the number after 12 is found to be 15. Further extending the pattern, the number are 18, 21, 24, 27. The pattern of increasing by multiples of three can describe this pattern.

A pattern can also be generated from a given rule. Starting with zero, the rule of adding 5 can be used to produce a set of numbers. The following list will result from using the rule: 0, 5, 10, 15, 20. Describing this pattern can include words such as "multiples" of 5 and an "increase" of 5. Any time this pattern needs to be extended, the rule can be applied to find more numbers. Patterns are identified by the rules they follow. This rule should be able to generate new numbers or shapes, while also applying to the given numbers or shapes.

Making Conjectures, Predictions, and Generalizations Based on Patterns
Given a certain pattern, future numbers or shapes can be found. Pascal's triangle is an example of a pattern of numbers. Questions can be asked of the triangle, such as, "what comes next?" and "what values determine the next line?" By examining the different parts of the triangle, conjectures can be made about how the numbers are generated. For the first few rows of numbers, the increase is small. Then the numbers begin to increase more quickly. By looking at each row, a conjecture can be made that the sum of the first row determines the second row's numbers. The second row's numbers can be added to find the third row. To test this conjecture, two numbers can be added, and the number found directly between and below them should be that sum. For the third row, the middle number is 2, which is the sum of the two 1s above it. For the fifth row, the 1 and 3 can be added to find a sum of 4, the same number below the 1 and 3. This pattern continues throughout the triangle. Once the pattern is confirmed to apply throughout the triangle, predictions can be made for future rows. The sums of the bottom row numbers can be found and then added to the bottom of the triangle. In more general terms, the diagonal rows have patterns as well. The outside numbers are always 1. The second diagonal rows are in counting order. The third diagonal row increases each time by one more than the previous. It is helpful to generalize patterns because it makes the pattern more useful in terms of applying it. Pascal's triangle can be used to

predict the tossing of a coin, predicting the chances of heads or tails for different numbers of tosses. It can also be used to show the Fibonacci Sequence, which is found by adding the diagonal numbers together.

Pascal's Triangle

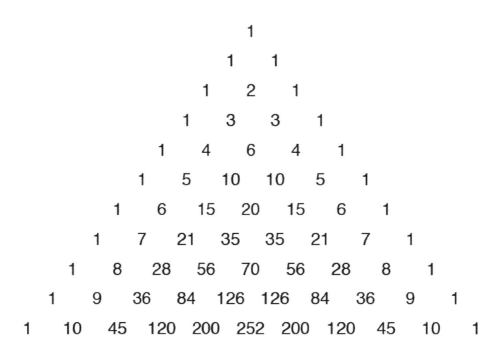

Identifying Relationships Between the Corresponding Terms of Two Numerical Patterns
Sets of numerical patterns can be found by starting with a number and following a given rule. If two sets are generated, the corresponding terms in each set can be found to relate to one another by one or more operations. For example, the following table shows two sets of numbers that each follow their own pattern. The first column shows a pattern of numbers increasing by 1. The second column shows the numbers increasing by 4. Because the numbers are lined up, corresponding numbers are side by side for the two sets. A question to ask is, "How can the number in the first column be turned into the number in the second column?"

1	4
2	8
3	12
4	16
5	20

This answer will lead to the relationship between the two sets. By recognizing the multiples of 4 in the right column and the counting numbers in order in the left column, the relationship of multiplying by four is determined. The first set is multiplied by 4 to get the second set of numbers. To confirm this relationship, each set of corresponding numbers can be checked. For any two sets of numerical patterns, the corresponding numbers can be lined up to find how each one relates to the other. In some cases, the relationship is simply addition or subtraction, multiplication or division. In other relationships, these operations are used in conjunction with each together. As seen in the following table, the relationship uses multiplication and addition. The following expression shows this relationship: $3x + 2$. The x represents the numbers in the first column.

1	5
2	8
3	11
4	14

Geometry and Measurement

Geometric Measurement and Dimension

Lines

Geometric figures can be identified by matching the definition with the object. For example, a **line segment** is made up of two endpoints and the line drawn between them. A **ray** is made up of one endpoint and one extending side that goes on forever. A **line** has no endpoints and two sides that extend on forever. These three geometric figures are shown below. What happens at A and B determines the name of each figure.

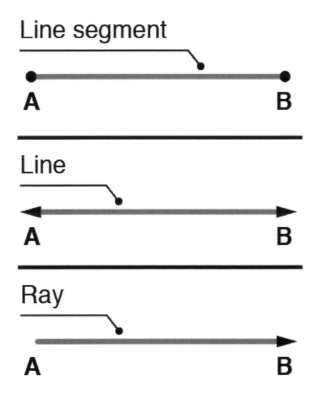

Parallel and perpendicular lines are made up of two lines, like the second figure above. They are distinguished from each other by how the two lines interact. **Parallel** lines run alongside one another, but they never intersect. **Perpendicular** lines intersect at a 90-degree, or a right, angle. An example of these two sets of lines is show below. Also shown in the figure are non-examples of these two types of lines.

Because the first set of lines, in the top left corner, will eventually intersect if they continue, they are not parallel. In the second set, the lines run in the same direction and will never intersect, making them parallel. The third set, in the bottom left corner, intersect at an angle that is not right, or not 90 degrees. The fourth set is perpendicular because it intersects at exactly a right angle.

Lines

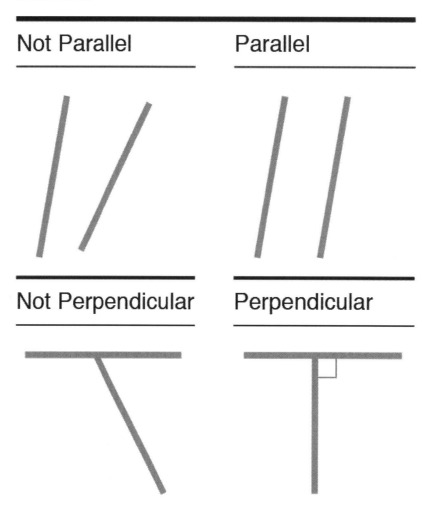

Not Parallel

Parallel

Not Perpendicular

Perpendicular

Angles

When two rays are joined together at their endpoints, an **angle** is formed. Angles can be described based on their measure. An angle whose measure is 90 degrees is described as a right angle, just as with perpendicular lines. Ninety degrees is a standard, to which other angles are compared. If an angle is less than ninety degrees, it is an **acute angle**. If it is greater than ninety degrees, it is an **obtuse angle**. If an angle is equal to twice a right angle, or 180 degrees, it is a **straight angle**.

Examples of these types of angles are shown below:

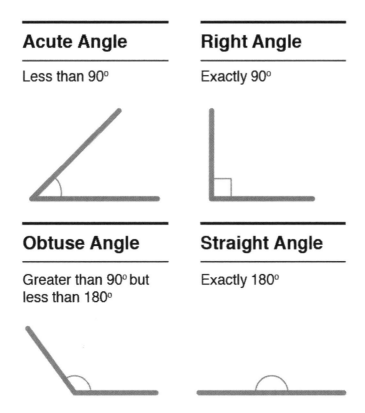

Acute Angle

Less than 90°

Right Angle

Exactly 90°

Obtuse Angle

Greater than 90° but less than 180°

Straight Angle

Exactly 180°

A **straight angle** is equal to 180 degrees, or a straight line. If the line continues through the **vertex,** or point where the rays meet, and does not change direction, then the angle is straight. This is shown in Figure 1 below. The second figure shows an obtuse angle. Its measure is greater than ninety degrees, but less than that of a straight angle. An estimate for its measure may be 175 degrees. Figure 3 shows an acute angle because it is just less than that of a right angle. Its measure may be estimated to be 80 degrees.

The last image, Figure 4, shows another acute angle. This measure is much smaller, at approximately 35 degrees, but it is still classified as acute because it is between zero and 90 degrees.

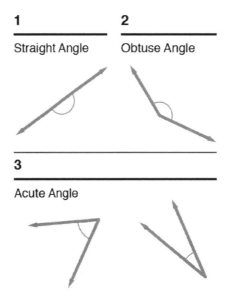

1

Straight Angle

2

Obtuse Angle

3

Acute Angle

Other geometric uses of angles can include the measure of angles inside a triangle. The sum of the measures of three angles in any triangle is 180 degrees. Therefore, if only two angles are known inside a triangle, the third can be found by subtracting the sum of the two known quantities from 180. Two angles whose sum is equal to 90 degrees are known as **complementary angles.** For example, angles measuring 72 and 18 degrees are complementary, and each angle is a complement of the other. Finally, two angles whose sum is equal to 180 degrees are known as **supplementary angles.** To find the supplement of an angle, subtract the given angle from 180 degrees. For example, the supplement of an angle that is 50 degrees is 180 – 50 = 130 degrees.

These terms involving angles can be seen in many types of word problems. For example, consider the following problem: The measure of an angle is 60 degrees less than two times the measure of its complement. What is the angle's measure? To solve this, let x be the unknown angle. Therefore, its complement is $90 - x$. The problem gives that:

$$x = 2(90 - x) - 60$$

To solve for x, distribute the 2, and collect like terms. This process results in:

$$x = 120 - 2x$$

Then, use the addition property to add $2x$ to both sides to obtain $3x = 120$. Finally, use the multiplication properties of equality to divide both sides by 3 to get $x = 40$. Therefore, the angle measures 40 degrees. Also, its complement measures 50 degrees.

Perimeter
Perimeter and area are two commonly used geometric quantities that describe objects. **Perimeter** is the distance around an object. The perimeter of an object can be found by adding the lengths of all sides. Perimeter may be used in problems dealing with lengths around objects such as fences or borders. It may also be used in finding missing lengths, or working backwards. If the perimeter is given, but a length is missing, subtraction can be used to find the missing length. Given a square with side length s, the formula

for perimeter is $P = 4s$. Given a rectangle with length l and width w, the formula for perimeter is $P = 2l + 2w$. The perimeter of a triangle is found by adding the three side lengths, and the perimeter of a trapezoid is found by adding the four side lengths. The units for perimeter are always the original units of length, such as meters, inches, miles, etc. When discussing a circle, the distance around the object is referred to as its **circumference,** not perimeter. The formula for circumference of a circle is $C = 2\pi r$, where r represents the radius of the circle. This formula can also be written as $C = d\pi$, where d represents the diameter of the circle.

Area
Area is the two-dimensional space covered by an object. These problems may include the area of a rectangle, a yard, or a wall to be painted. Finding the area may be a simple formula, or it may require multiple formulas to be used together. The units for area are square units, such as square meters, square inches, and square miles. Given a square with side length s, the formula for its area is $A = s^2$. Some other common shapes are shown below:

Shape	Formula	Graphic
Rectangle	$Area = length \times width$	
Triangle	$Area = \frac{1}{2} \times base \times height$	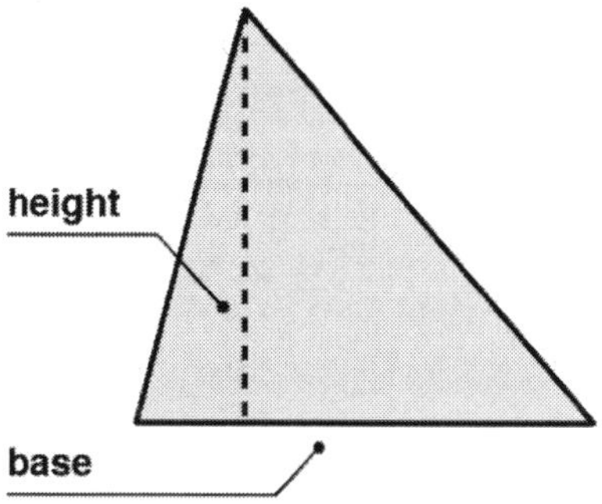
Circle	$Area = \pi \times radius^2$	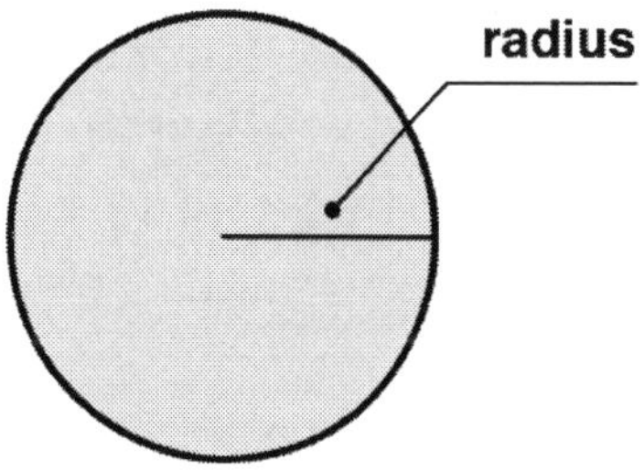

The following formula, not as widely used as those shown above, but very important, is the area of a trapezoid:

Area of a Trapezoid

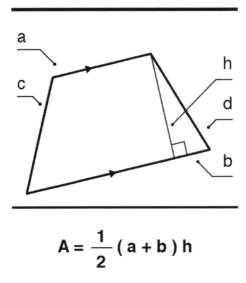

$$A = \frac{1}{2}(a+b)h$$

To find the area of the shapes above, use the given dimensions of the shape in the formula. Complex shapes might require more than one formula. To find the area of the figure below, break the figure into two shapes. The rectangle has dimensions 6 cm by 7 cm. The triangle has dimensions 6 cm by 6 cm. Plug the dimensions into the rectangle formula: $A = 6 \times 7$. Multiplication yields an area of 42 cm². The triangle area can be found using the formula $A = \frac{1}{2} \times 4 \times 6$. Multiplication yields an area of 12 cm². Add the areas of the two shapes to find the total area of the figure, which is 54 cm².

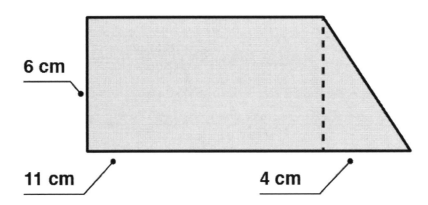

Instead of combining areas, some problems may require subtracting them, or finding the difference.

To find the area of the shaded region in the figure below, determine the area of the whole figure. Then the area of the circle can be subtracted from the whole.

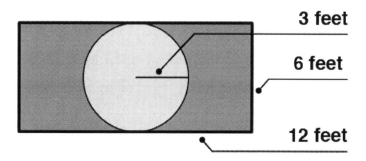

The following formula shows the area of the outside rectangle: $A = 12 \times 6 = 72\ ft^2$. The area of the inside circle can be found by the following formula: $A = \pi(3)^2 = 9\pi = 28.3\ ft^2$. As the shaded area is outside the circle, the area for the circle can be subtracted from the area of the rectangle to yield an area of $43.7\ ft^2$.

While some geometric figures may be given as pictures, others may be described in words. If a rectangular playing field with dimensions 95 meters long by 50 meters wide is measured for perimeter, the distance around the field must be found. The perimeter includes two lengths and two widths to measure the entire outside of the field. This quantity can be calculated using the following equation: $P = 2(95) + 2(50) = 290\ m$. The distance around the field is 290 meters.

Volume
Perimeter and area are two-dimensional descriptions; volume is three-dimensional. **Volume** describes the amount of space that an object occupies, but it's different from area because it has three dimensions instead of two. The units for volume are cubic units, such as cubic meters, cubic inches, and cubic millimeters. Volume can be found by using formulas for common objects such as cylinders and boxes.

The following chart shows a diagram and formula for the volume of two objects:

Shape	Formula	Diagram
Rectangular Prism (box)	$V = length \times width \times height$	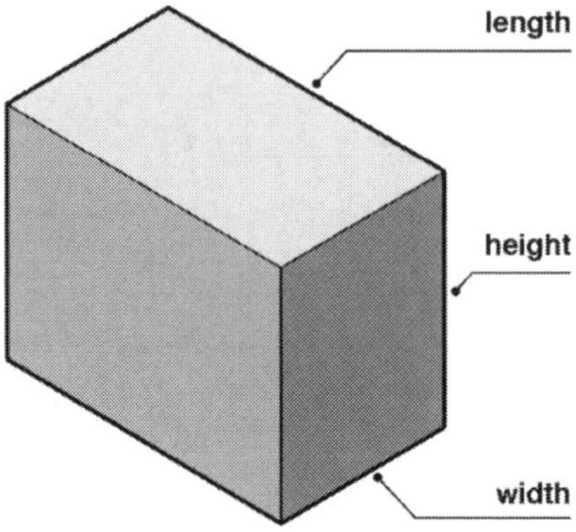
Cylinder	$V = \pi \times radius^2 \times height$	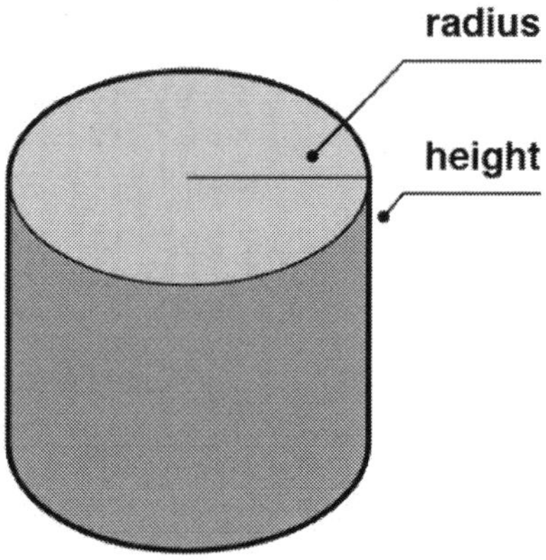

Volume formulas of these two objects are derived by finding the area of the bottom two-dimensional shape, such as the circle or rectangle, and then multiplying times the height of the three-dimensional shape. Other volume formulas include the volume of a cube with side length s: $V = s^3$; the volume of a sphere with radius r. $V = \frac{4}{3}\pi r^3$; and the volume of a cone with radius r and height h.

$$V = \frac{1}{3}\pi r^2 h$$

If a soda can has a height of 5 inches and a radius on the top of 1.5 inches, the volume can be found using one of the given formulas. A soda can is a cylinder. Knowing the given dimensions, the formula can be completed as follows:

$$V = \pi \, (radius)^2 \times height$$

$$\pi \, (1.5)^2 \times 5 = 35.325 \ inches^3$$

Notice that the units for volume are inches cubed because it refers to the number of cubic inches required to fill the can.

Right rectangular prisms are those prisms in which all sides are rectangles and all angles are right, or equal to 90 degrees. The volume for these objects can be found by multiplying the length by the width by the height. The formula is $V = lwh$. For the following prism, the volume formula is:

$$V = 6\frac{1}{2} \times 3 \times 9$$

When dealing with fractional edge lengths, it is helpful to convert the length to an improper fraction. The length 6 ½ cm becomes $\frac{13}{2}$ cm. Then the formula becomes:

$$V = \frac{13}{2} \times 3 \times 9$$

$$\frac{13}{2} \times \frac{3}{1} \times \frac{9}{1} = \frac{351}{2}$$

This value for volume is better understood when turned into a mixed number, which would be 175 ½ cm³.

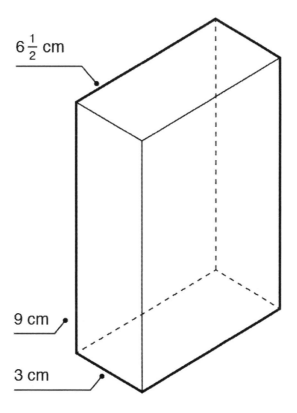

When dimensions for length are given with fractional parts, it can be helpful to turn the mixed number into an improper fraction, then multiply to find the volume, then convert back to a mixed number.

Surface Area

Surface area is defined as the area of the surface of a figure. A **pyramid** has a surface made up of four triangles and one square. To calculate the surface area of a pyramid, the areas of each individual shape are calculated. Then the areas are added together. This method of decomposing the shape into two-dimensional figures to find area, then adding the areas, can be used to find surface area for any figure. Once these measurements are found, the area is described with square units. For example, the following figure shows a rectangular prism. The figure beside it shows the rectangular prism broken down into two-dimensional shapes, or rectangles. The area of each rectangle can be calculated by multiplying the length by the width. The area for the six rectangles can be represented by the following expression:

$$5 \times 6 + 5 \times 10 + 5 \times 6 + 6 \times 10 + 5 \times 10 + 6 \times 10$$

The total for all these areas added together is $280m^2$, or 280 square meters. This measurement represents the surface area because it is the area of all six surfaces of the rectangular prism.

The Net of a Rectangular Prism

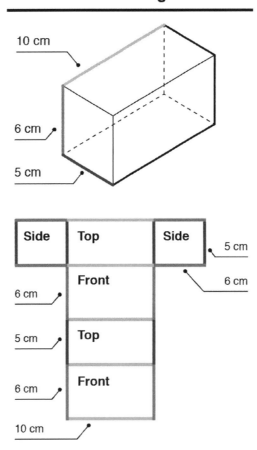

Another shape that has a surface area is a cylinder. The shapes used to make up the **cylinder** are two circles and a rectangle wrapped around between the two circles. A common example of a cylinder is a can. The two circles that make up the bases are obvious shapes. The rectangle can be more difficult to see, but the label on a can will help illustrate it. When the label is removed from a can and laid flat, the shape is a rectangle. When the areas for each shape are needed, there will be two formulas. The first is the area for the circles on the bases. This area is given by the formula $A = \pi r^2$. There will be two of these areas—one for the top and one for the bottom if the can (cylinder) is standing upright on a shelf. Then the area of the rectangle must be determined. The width of the rectangle is equal to the height of the can, *h*. The length of the rectangle is equal to the circumference of the base circle, $2\pi r$. The area for the rectangle can be found by using the formula $A = 2\pi r \times h$. By adding the two areas for the bases and the area of the rectangle, the surface area of the cylinder can be found, described in units squared.

The surface area of a rectangular prism can be found by breaking down the figure into basic shapes. These shapes are rectangles, made up of the two bases, two sides, and the front and back. Consider the rectangular prism we previously looked at:

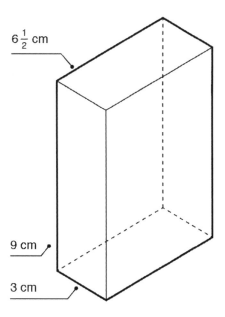

The formula for the surface area uses the area for each of these shapes for the terms in the following equation:

$$SA = 6\frac{1}{2} \times 3 + 6\frac{1}{2} \times 3 + 3 \times 9 + 3 \times 9 + 6\frac{1}{2} \times 9 + 6\frac{1}{2} \times 9$$

Because there are so many terms in a surface area formula and because this formula contains a fraction, it can be simplified by combining groups that are the same. Each set of numbers is used twice, to represent areas for the opposite sides of the prism. The formula can be simplified to:

$$SA = 2\left(6\frac{1}{2} \times 3\right) + 2(3 \times 9) + 2\left(6\frac{1}{2} \times 9\right)$$

$$2\left(\frac{13}{2} \times 3\right) + 2(27) + 2\left(\frac{13}{2} \times 9\right)$$

$$2\left(\frac{39}{2}\right) + 54 + 2\left(\frac{117}{2}\right)$$

$$39 + 54 + 117 = 210 \text{ cm}^2$$

Using Nets to Determine the Surface Area of Three-Dimensional Figures
The **net of a figure** is the resulting two-dimensional shapes when a three-dimensional shape is broken down. Nets can be used in calculating different values for given shapes. One useful way to calculate

surface area is to find the net of the object, then find the area of each of the shapes and add them together. Nets are also useful in composing shapes and decomposing objects so as to view how objects connect and can be used in conjunction with each other.

Surface area of three-dimensional figures is the total area of each of the faces of the figures. Nets are used to lay out each face of an object. The following figure shows a triangular prism. The bases are triangles and the sides are rectangles. The second figure shows the net for this triangular prism. The dimensions are labeled for each of the faces of the prism. The area for each of the two triangles can be determined by the formula:

$$A = \frac{1}{2}bh = \frac{1}{2} \times 8 \times 9 = 36cm^2$$

The rectangle areas can be described by the equation:

$$A = lw = 8 \times 5 + 9 \times 5 + 10 \times 5$$

$$40 + 45 + 50 = 135cm^2$$

The area for the triangles can be multiplied by two, then added to the rectangle areas to yield a total surface are of $207cm^2$.

A Triangular Prism and Its Net

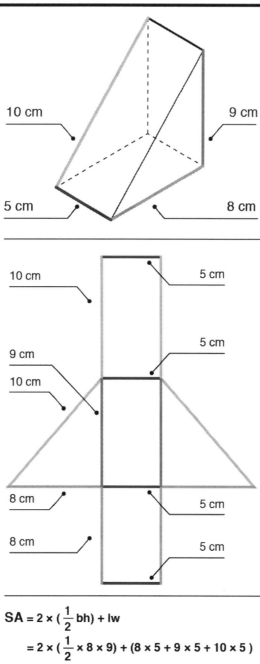

$$SA = 2 \times (\frac{1}{2}bh) + lw$$

$$= 2 \times (\frac{1}{2} \times 8 \times 9) + (8 \times 5 + 9 \times 5 + 10 \times 5)$$

$$= 207cm^2$$

Other figures that have rectangles or triangles in their nets include pyramids, rectangular prisms, and cylinders. When the shapes of these three-dimensional objects are found, and areas are calculated, the sum will result in the surface area. The following picture shows the net for a rectangular prism. The dimensions for each of the shapes that make up the prism are shown to the right. As a formula, the surface area is the sum of each rectangle added together. The following equation shows the formula:

$$SA = 5 \times 10 + 5 \times 6 + 6 \times 10 + 5 \times 6 + 5 \times 10 + 6 \times 10$$

$$50 + 30 + 60 + 30 + 50 + 60 = 280m^2$$

A Rectangular Prism and its Net

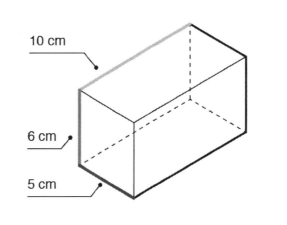

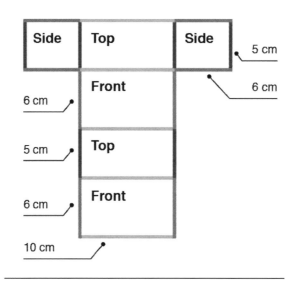

SA = 5×10 + 5×6 + 6×10 + 5×6 + 5×10 + 6×10

= 50 + 30 + 60 + 30 + 50 + 60

= 280cm²

A net for a pyramid is shown in the figure below. The base of the pyramid is a square. The four shapes coming off the pyramid are triangles. When built up together, folding the triangles to the top results in a pyramid. Surface area of the pyramid could be calculated by summing the area of each of the constituent shapes in the net.

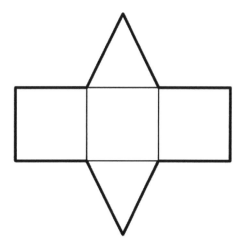

Another example of a net of a figure, in this case a cylinder, is shown below. When the cylinder is broken down, the bases are circles and the side is a rectangle wrapped around the circles. The circumference of the circle turns into the length of the rectangle.

The circumference
of the circle is
equal to the length
of the rectangle

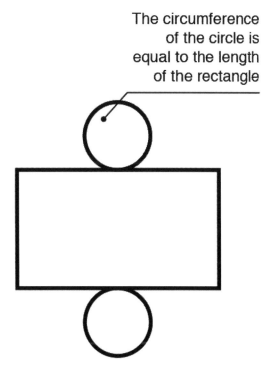

Determining How Changes to Dimensions Change Area and Volume

When the dimensions of an object change, the area and volume are also subject to change. For example, the following rectangle has an area of 98 square centimeters ($A = 7 \times 14 = 98\text{cm}^2$). If the length is increased by 2, to be 16 cm, then the area becomes:

$$A = 7 \times 16 = 112\text{cm}^2$$

The area increased by 14 cm, or twice the width because there were two more widths of 7 cm.

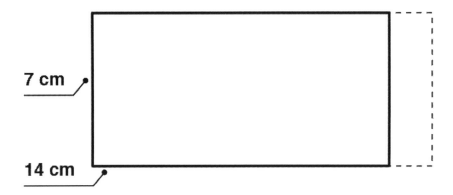

For the volume of an object, there are three dimensions. The given prism has a volume of:

$$V = 4 \times 12 \times 3 = 144\text{m}^2$$

If the height is increased by 3, the volume becomes:

$$V = 4 \times 12 \times 6 = 288\text{m}^2$$

The increase of 3 for the height, or doubling of the height, resulted in a volume that was doubled. From the original, if the width was doubled, the volume would be:

$$V = 8 \times 12 \times 3 = 288\text{m}^2$$

When the width doubled, the volume was doubled also. The same increase in volume would result if the length was doubled.

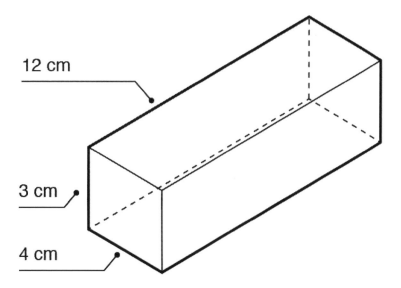

Solving for Missing Values in Triangles, Circles, and Other Figures

Solving for missing values in shapes requires knowledge of the shape and its characteristics. For example, a triangle has three sides and three angles that add up to 180 degrees. If two angle measurements are given, the third can be calculated. For the triangle below, the one given angle has a measure of 55 degrees. The missing angle is x. The third angle is labeled with a square, which indicates a measure of 90 degrees. Because all angles must sum to 180 degrees, the following equation can be used to find the missing x-value:

$$55 + 90 + x = 180$$

Adding the two given angles and subtracting the total from 180, the missing angle is found to be 35 degrees.

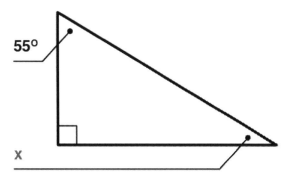

A similar problem can be solved with circles. If the radius is given, but the circumference is unknown, it can be calculated based on the formula $C = 2\pi r$. This example can be used in the figure below. The radius can be substituted for r in the formula.

Then the circumference can be found as:

$$C = 2\pi \times 8 = 16\pi = 50.24 \; cm$$

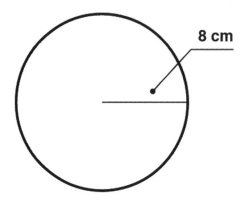

Other figures that may have missing values could be the length of a square, given the area, or the perimeter of a rectangle, given the length and width. All of the missing values can be found by first identifying all the characteristics that are known about the shape, then looking for ways to connect the missing value to the given information.

With any geometric calculations, it's important to determine what dimensions are given and what quantities the problem is asking for. If a connection can be made between them, the answer can be found.

Shapes and Solids

Shapes are defined by their angles and number of sides. A shape with one continuous side, where all points on that side are equidistant from a center point is called a **circle.** A shape made with three straight line segments is a **triangle.** A shape with four sides is called a **quadrilateral,** but more specifically a *square,* **rectangle, parallelogram**, or **trapezoid,** depending on the interior angles. These shapes are two-dimensional and only made of straight lines and angles.

Solids can be formed by combining these shapes and forming three-dimensional figures. These figures have another dimension because they add one more direction. Examples of solids may be prisms or spheres. There are four figures below that can be described based on their sides and dimensions. Figure 1 is a cone because it has three dimensions, where the bottom is a circle and the top is formed by the sides combining to one point. Figure 2 is a triangle because it has two dimensions, made up of three line segments. Figure 3 is a cylinder made up of two base circles and a rectangle to connect them in three dimensions. Figure 4 is an oval because it is one continuous line in two dimensions, not equidistant from the center.

Shapes and Solids

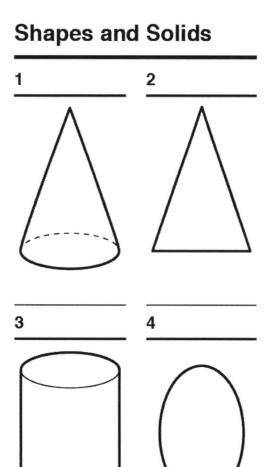

1

2

3

4

Figure 5 below is made up of squares in three dimensions, combined to make a cube. Figure 6 is a rectangle because it has four sides that intersect at right angles. More specifically, it can be described as a **square** because the four sides have equal measures. Figure 7 is a pyramid because the bottom shape is a square and the sides are all triangles. These triangles intersect at a point above the square. Figure 8 is a circle because it is made up of one continuous line where the points are all equidistant from one center point.

Shapes and Solids

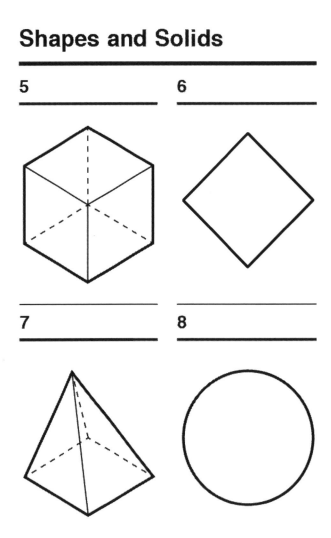

5

6

7

8

Basic shapes are those polygons that are made up of straight lines and angles and can be described by their number of sides and concavity. Some examples of those shapes are rectangles, triangles, hexagons, and pentagons. These shapes have identifying characteristics on their own, but they can also be decomposed into other shapes. For example, the following can be described as one hexagon, as see in the first figure. It can also be decomposed into six equilateral triangles. The last figure shows how the hexagon can be decomposed into three rhombuses.

Decomposing a Hexagon

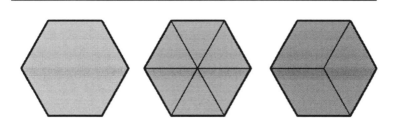

More complex shapes can be formed by combining basic shapes, or lining them up side by side. Below is an example of a house. This house is one figure all together, but can be decomposed into seven different shapes. The chimney is a parallelogram and the roof is made up of two triangles. The bottom of the house is a square alongside three triangles. There are many other ways of decomposing this house. Different shapes can be used to line up together and form one larger shape. The area for the house can be calculated by finding the individual areas for each shape, then adding them all together. For this house, there would be the area of four triangles, one square, and one parallelogram. Adding these all together would result in the area of the house as a whole. Decomposing and composing shapes is commonly done with a set of tangrams. A **tangram** is a set of shapes that includes different size triangles, rectangles, and parallelograms.

A Tangram of a House

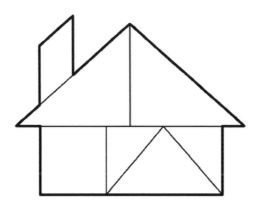

67

Congruence and Similarity

Two figures are **congruent** if they have the same shape and same size, meaning same angle measurements and equal side lengths. Two figures are **similar** if they have the same angle measurement but not side lengths. Basically, angles are congruent in similar triangles and their side lengths are constant multiplies of each other. Proving two shapes are similar involves showing that all angles are the same; proving two shapes are congruent involves showing that all angles are the same *and* that all sides are the same. If two pairs of angles are congruent in two triangles, then those triangles are similar because their third angle has to be equal due to the fact that all three angles add up to 180 degrees.

There are five main theorems that are used to show triangles are congruent. Each theorem involves showing different combinations of sides and angles are the same in two triangles, which proves the triangles are congruent. The **side-side-side (SSS) theorem** states that if all sides are equal in two triangles, the triangles are congruent. The **side-angle-side (SAS) theorem** states that if two pairs of sides are equal and the included angles are congruent in two triangles, then the triangles are congruent.

Similarly, the **angle-side-angle (ASA) theorem** states that if two pairs of angles are congruent and the included side lengths are equal in two triangles, the triangles are similar. The **angle-angle-side (AAS) theorem** states that two triangles are congruent if they have two pairs of congruent angles and a pair of corresponding equal side lengths that are not included. Finally, the **hypotenuse-leg (HL) theorem** states that if two right triangles have equal hypotenuses and an equal pair of shorter sides, the triangles are congruent. An important item to note is that **angle-angle-angle (AAA)** is not enough information to have congruence because if three angles are equal in two triangles, the triangles can only be described as similar.

The Pythagorean Theorem and Right Triangles

Within right triangles, trigonometric ratios can be defined for the acute angle within the triangle. Consider the following right triangle. The side across from the right angle is known as the **hypotenuse,** the acute angle being discussed is labeled θ, the side across from the acute angle is known as the **opposite** side, and the other side is known as the **adjacent** side.

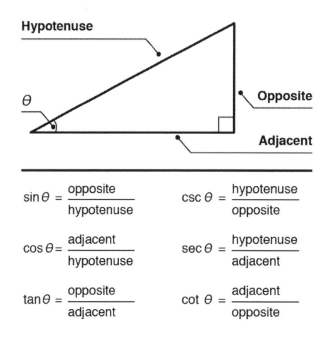

$$\sin\theta = \frac{\text{opposite}}{\text{hypotenuse}} \qquad \csc\theta = \frac{\text{hypotenuse}}{\text{opposite}}$$

$$\cos\theta = \frac{\text{adjacent}}{\text{hypotenuse}} \qquad \sec\theta = \frac{\text{hypotenuse}}{\text{adjacent}}$$

$$\tan\theta = \frac{\text{opposite}}{\text{adjacent}} \qquad \cot\theta = \frac{\text{adjacent}}{\text{opposite}}$$

The six trigonometric ratios are shown above as well. "Sin" is short for sine, "cos" is short for cosine, "tan" is short for tangent, "csc" is short for cosecant, "sec" is short for secant, and "cot" is short for cotangent. A mnemonic device exists that is helpful to remember the ratios. SOHCAHTOH stands for Sine = Opposite/Hypotenuse, Cosine = Adjacent/Hypotenuse, and Tangent = Opposite/Adjacent. The other three trigonometric ratios are reciprocals of sine, cosine, and tangent because $\csc \theta = \frac{1}{\sin \theta}$, $\sec \theta = \frac{1}{\cos \theta}$, and $\cot \theta = \frac{1}{\tan \theta}$.

The **Pythagorean Theorem** is an important relationship between the three sides of a right triangle. It states that the square of hypotenuse is equal to the sum of the squares of the other two sides. When using the Pythagorean Theorem, the hypotenuse is labeled as side c, the opposite is labeled as side a, and the adjacent side is side b.

The theorem can be seen in the following diagram:

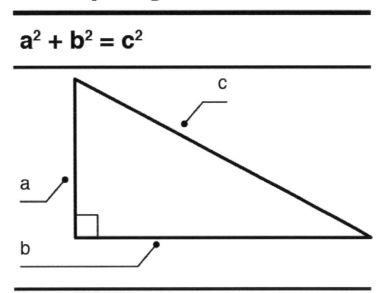

Both the trigonometric ratios and Pythagorean Theorem can be used in problems that involve finding either a missing side or missing angle of a right triangle. Look to see what sides and angles are given and select the correct relationship that will assist in finding the missing value. These relationships can also be used to solve application problems involving right triangles. Often, it is helpful to draw a figure to represent the problem to see what is missing.

Circles

The formula for area of a circle is $A = \pi r^2$ and therefore, formula for area of a **sector** is $\pi r^2 \frac{A}{360}$, a fraction of the entire area of the circle. If the radius of a circle and arc length is known, the central angle measurement in degrees can be found by using the formula:

$$\frac{360 \cdot arclength}{2\pi r}$$

If the desired central angle measurement is in radians, the formula for the central angle measurement is much simpler as $\frac{arc\ length}{r}$.

The Center, Radius, Central Angle, a Sector, and an Arc of a Circle

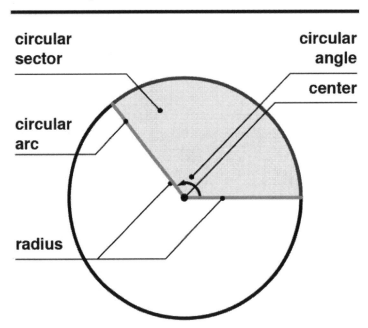

70

A **chord** of a circle is a straight-line segment that connects any two points on a circle. The line segment does not have to travel through the center, as the diameter does. Also, note that the chord stops at the circumference of the circle. If it did not stop and extended toward infinity, it would be known as a **secant line.** The following shows a diagram of a circle with a chord shown by the dotted line. The radius is r and the central angle is A:

A Circle with a Chord

Chord Length $= 2\, r \sin\dfrac{A}{2}$

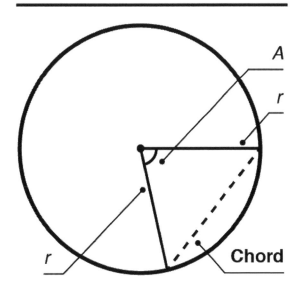

One formula for chord length can be seen in the diagram, and is equal to $2r \sin\frac{A}{2}$, where A is the central angle. Another formula for chord length is: chord length $= 2\sqrt{r^2 - D^2}$, where D is equal to the distance from the chord to the center of the circle. This formula is basically a version of the Pythagorean Theorem.

Formulas for chord lengths vary based on what type of information is known. If the radius and central angle are known, the first formula listed above should be used by plugging the radius and angle in directly. If the radius and the distance from the center to the chord are known, the second formula listed previously (chord length $= 2\sqrt{r^2 - D^2}$) should be used.

Many theorems exist between arc lengths, angle measures, chord lengths, and areas of sectors. For instance, when two chords intersect in a circle, the product of the lengths of the individual line segments are equal. For instance, in the following diagram, $A \times B = C \times D$.

$$A \times B = C \times D$$

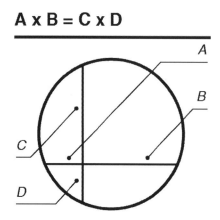

Plane Geometry

The Coordinate Plane

The coordinate plane is a way of identifying the position of a point in relation to two axes. The **coordinate plane** is made up of two intersecting lines, the x-axis and the y-axis. These lines intersect at a right angle, and their intersection point is called the **origin**. The points on the coordinate plane are labeled based on their position in relation to the origin. If a point is found 4 units to the right and 2 units up from the origin, the location is described as (4, 2). These numbers are the x- and y-coordinates, always written in the order (x, y). This point is also described as lying in the first quadrant. Every point in the first quadrant has a location that is positive in the x and y directions.

The following figure shows the coordinate plane with examples of points that lie in each quadrant.

The Coordinate Plane

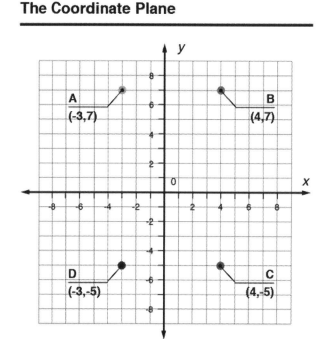

Point B lies in the first quadrant, described with positive x- and y-values, above the x-axis and to the right of the y-axis. Point A lies in the second quadrant, where the x-value is negative and the y-value is positive. This quadrant is above the x-axis and to the left of the y-axis. Point D lies in the third quadrant, where both the x- and y-values are negative. Points in this quadrant are described as being below the x-axis and to the left of the y-axis. Point C is in the fourth quadrant, where the x-value is positive and the y-value is negative.

Circles in the Coordinate Plane
Recall that a **circle** is the set of all points the same distance, known as radius r, from a single point C, known as the center of the circle. The center has coordinates (h, k) and any point on the circle is an ordered pair with coordinates (x, y). A right triangle with hypotenuse r can be formed with these two points as seen here:

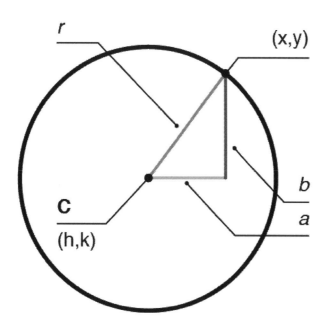

The other side lengths are a and b. The **Pythagorean Theorem** states that $a^2 + b^2 = r^2$. However, a can be replaced by $|x - h|$ and b can be replaced by $|y - k|$ because the distance between any two coordinates in the coordinate plane is the absolute value of their difference. That substitution gives $(x - h)^2 + (y - k)^2 = r^2$, which is the formula used to find the equation of any circle with center (h, k) and radius r. Therefore, if any problem gives the coordinates of the center of a circle and its radius length, this is the equation in two variables that allows any other point on the circle to be found.

Oftentimes, the center or the radius of a circle are not easily seen in the given equation of the circle. If the equation is in standard form of a polynomial equation like $ax^2 + ay^2 + cx + dy + e = 0$, the algebraic technique of completing the square must be used to find the coordinates of the center and the radius. Completing the square must be done within both variables x and y. First, the constant term needs to be subtracted off of both sides of the equation, and then the x and y terms need to be grouped together. Then, the entire equation needs to be divided by a. Then, divide the coefficient of the x-term by 2, square it, and add that value to both sides of the equation. This value should be grouped with the x terms. Next, divide the coefficient of the y-term by 2, square it, and add it to both sides of the equation, grouping it with the y terms. The trinomial in both x and y are now perfect square trinomials and can be factored into squares of a binomial. This process results in $(x - h)^2 + (y - k)^2 = r^2$, showing the radius and coordinates of the center.

Transformations in the Plane
Two-dimensional figures can undergo various types of transformations in the plane. They can be translated (shifted) horizontally and vertically, reflected, compressed, or stretched.

A **translation,** also known as a **shift** or **slide,** moves the shape in one direction. Here is a picture of a translation:

A Translation

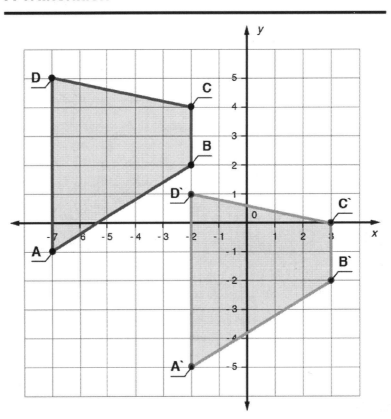

Notice that the size of the original shape has not changed at all. If a translation exists within a shape drawn on the Cartesian coordinate system, the translation can be represented by adding or subtracting values onto the x-, y-, or both the x- and y-coordinates of the points. However, all vertices will move the same number of units because the shape and size of the shape do not change.

A figure can also be **reflected** or flipped, and this transformation involves reflecting over a given line, known as the **line of reflection.** The original triangle (called the **preimage**) is seen in the figure below in the first quadrant. The reflection of this triangle is in the third quadrant. A reflection across the y-axis can be found by determining each point's distance to the y-axis and moving it that same distance on the opposite side. For example, the point C is located at (4, 1). The reflection of this point moves to (–4, 1) when reflected across the y-axis. The original point A is located at (1, 3), and the reflection across the y-

axis is located at (–1, 5). It is evident that the reflection across the *y*-axis changes the sign of the *x*-coordinate. A reflection across the *x*-axis changes the sign of the *y*-coordinate instead.

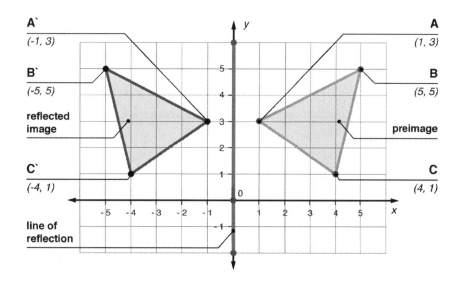

Similarly, if the shape is reflected over the *x*-axis, the *y*-coordinate stays the same, but the *x*-coordinates are made negative. For instance, in the graphic below, the point C at (3, 5) becomes C' at (3, -5).

A Reflection Over the X-Axis

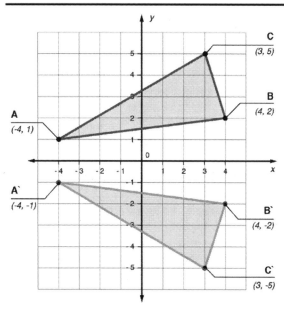

Thirdly, a **compression** or a **stretch** of a figure involves changing the size of the original figure, and together they can be classified as a **dilation**. A **compression** shrinks the size of the figure. We can think about this as a multiplication process by multiplying times a value between 0 and 1. A **stretch** of a figure results in a figure larger than the original shape. If we consider multiplication, the factor would be greater than 1.

Here is a picture of a dilation that is comprised of a stretch in which the original square doubled in size:

A Dilation with a Scale Factor of 2

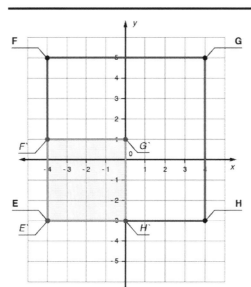

If a shape within the Cartesian coordinate system gets stretched, its coordinates get multiplied by a number greater than 1, and if a shape gets compressed, its coordinates get multiplied by a number between 0 and 1.

A figure can undergo any combination of transformations. For instance, it can be shifted, reflected, and stretched at the same time.

Converting Within and Between Standard and Metric Systems

When working with dimensions, sometimes the given units don't match the formula, and conversions must be made. The metric system has base units of meter for length, kilogram for mass, and liter for liquid volume. This system expands to three places above the base unit and three places below. These places correspond with prefixes with a base of 10.

The following table shows the conversions:

kilo-	hecto-	deka-	base	deci-	centi-	milli-
1,000 times the base	100 times the base	10 times the base		1/10 times the base	1/100 times the base	1/1000 times the base

To convert between units within the metric system, values with a base ten can be multiplied. The decimal can also be moved in the direction of the new unit by the same number of zeros on the number. For example, 3 meters is equivalent to .003 kilometers. The decimal moved three places (the same number of zeros for kilo-) to the left (the same direction from base to kilo-). Three meters is also equivalent to 3,000 millimeters. The decimal is moved three places to the right because the prefix milli- is three places to the right of the base unit.

The English Standard system used in the United States has a base unit of foot for length, pound for weight, and gallon for liquid volume. These conversions aren't as easy as the metric system because they aren't a base ten model. The following table shows the conversions within this system.

Length	Weight	Capacity
1 foot (ft) = 12 inches (in) 1 yard (yd) = 3 feet 1 mile (mi) = 5280 feet 1 mile = 1760 yards	1 pound (lb) = 16 ounces (oz) 1 ton = 2000 pounds	1 tablespoon (tbsp) = 3 teaspoons (tsp) 1 cup (c) = 16 tablespoons 1 cup = 8 fluid ounces (oz) 1 pint (pt) = 2 cups 1 quart (qt) = 2 pints 1 gallon (gal) = 4 quarts

When converting within the English Standard system, most calculations include a conversion to the base unit and then another to the desired unit. For example, take the following problem: 3 $quarts$ = ___ $cups$. There is no straight conversion from quarts to cups, so the first conversion is from quarts to pints. There are 2 pints in 1 quart, so there are 6 pints in 3 quarts. This conversion can be solved as a proportion: $\frac{3\ qt}{x} = \frac{1\ qt}{2\ pints}$. It can also be observed as a ratio 2:1, expanded to 6:3. Then the 6 pints must be converted to cups. The ratio of pints to cups is 1:2, so the expanded ratio is 6:12. For 6 pints, the measurement is 12 cups. This problem can also be set up as one set of fractions to cancel out units. It begins with the given information and cancels out matching units on top and bottom to yield the answer. Consider the following expression:

$$\frac{3\ quarts}{1} \times \frac{2\ pints}{1\ quart} \times \frac{2\ cups}{1\ pint}$$

It's set up so that units on the top and bottom cancel each other out:

$$\frac{3\ \cancel{quarts}}{1} \times \frac{2\ \cancel{pints}}{1\ \cancel{quart}} \times \frac{2\ cups}{1\ \cancel{pint}}$$

The numbers can be calculated as 3 × 3 × 2 on the top and 1 on the bottom. It still yields an answer of 12 cups.

This process of setting up fractions and canceling out matching units can be used to convert between standard and metric systems. A few common equivalent conversions are 2.54 cm = 1 inch, 3.28 feet = 1 meter, and 2.205 pounds = 1 kilogram. Writing these as fractions allows them to be used in conversions. For the fill-in-the-blank problem 5 meters = ___ feet, an expression using conversions starts with the expression $\frac{5\ meters}{1} \times \frac{3.28\ feet}{1\ meter}$, where the units of meters will cancel each other out, and the final unit is feet. Calculating the numbers yields 16.4 feet. This problem only required two fractions. Others may require longer expressions, but the underlying rule stays the same. When there's a unit on the top of the fraction that's the same as the unit on the bottom, then they cancel each other out. Using this logic and the conversions given above, many units can be converted between and within the different systems.

The conversion between Fahrenheit and Celsius is found in a formula:

$$°C = (°F - 32) \times \frac{5}{9}$$

For example, to convert 78°F to Celsius, the given temperature would be entered into the formula:

$$°C = (78 - 32) \times \frac{5}{9}$$

Solving the equation, the temperature comes out to be 25.56°C. To convert in the other direction, the formula becomes:

$$°F = °C \times \frac{9}{5} + 32$$

Remember the order of operations when calculating these conversions.

Basic Statistics and Data Analysis

Basic Statistics

Statistical Variability
Statistics is the branch of mathematics that deals with the collection, organization, and analysis of data. A statistical question is one that can be answered by collecting and analyzing data. When collecting data, expect variability. For example, "How many pets does Yanni own?" is not a statistical question because it can be answered in one way. "How many pets do the people in a certain neighborhood own?" is a statistical question because, to determine this answer, one would need to collect data from each person in the neighborhood, and it is reasonable to expect the answers to vary.

Identify these as questions statistical or not statistical:

1. How old are you?

2. What is the average age of the people in your class?

3. How tall are the students in Mrs. Jones' sixth grade class?

4. Do you like Brussels sprouts?

Questions 1 and 4 are not statistical questions, but questions 2 and 3 are.

Data collection can be done through surveys, experiments, observations, and interviews. A **census** is a type of survey that is done with a whole population. Because it can be difficult to collect data for an entire population, sometimes a **sample** is used. In this case, one would survey only a fraction of the population and make inferences about the data. Sample surveys are not as accurate as a census, but it is an easier and less expensive method of collecting data. An **experiment** is used when a researcher wants to explain how one variable causes changes in another variable. For example, if a researcher wanted to know if a particular drug affects weight loss, he or she would choose a **treatment group** that would take the drug, and another group, the **control group**, that would not take the drug. Special care must be taken when choosing these groups to ensure that bias is not a factor. **Bias** occurs when an outside factor influences the outcome of the research. In observational studies, the researcher does not try to influence either variable but simply observes the behavior of the subjects. Interviews are sometimes used to collect data as well. The researcher will ask questions that focus on her area of interest in order to gain insight from the participants. When gathering data through observation or interviews, it is important that the researcher is well trained so that he or she does not influence the results and so that the study is reliable. A study is reliable if it can be repeated under the same conditions and the same results are received each time.

In statistics, a **population** contains all subjects being studied. For example, a population could be every student at a university or all males in the United States. A **sample** consists of a group of subjects from an entire population. A sample would be 100 students at a university or 100,000 males in the United States. **Inferential statistics** is the process of using a sample to generalize information concerning populations. **Hypothesis testing** is the actual process used when evaluating claims made about a population based on a sample.

A **statistic** is a measure obtained from a sample, and a **parameter** is a measure obtained from a population. For example, the mean SAT score of the 100 students in the sample at a university would be a statistic, and the mean SAT score of all university students would be a parameter.

The beginning stages of hypothesis testing starts with formulating a **hypothesis,** a statement made concerning a population parameter. The hypothesis may be true, or it may not be true. The experiment will help answer that question. In each setting, there are two different types of hypotheses: the **null hypothesis**, written as H_0, and the **alternative hypothesis**, written as H_1. The null hypothesis represents verbally when there is not a difference between two parameters, and the alternative hypothesis represents verbally when there is a difference between two parameters. Consider the following experiment: A researcher wants to see if a new brand of allergy medication has any effect on drowsiness of the patients who take the medication. He wants to know if the average hours spent sleeping per day increases. The mean for the population under study is 8 hours, so $\mu = 8$. In other words, the population parameter is μ, the mean. The null hypothesis is $\mu = 8$ and the alternative hypothesis is $\mu > 8$. When using a smaller sample of a population, the null hypothesis represents the situation when the mean remains unaffected and the alternative hypothesis represents the situation when the mean increases. The chosen statistical test will apply the data from the sample to actually decide whether the null hypothesis should or should not be rejected.

The Random Processes Underlying Statistical Experiments
For researchers to make valid conclusions about population characteristics and parameters, the sample used to compare must be random. In a **random sample**, every member of the population must have an equal chance of being selected. In this situation, the sample is **unbiased** and is said to be a good representation of the population. If a sample is selected in an inappropriate manner, it is said to be **biased.** A sample can be biased if, for example, some subjects were more likely to be chosen than others. In order to have unbiased samples, the four main sampling methods used tend to be random, systematic, stratified, and cluster sampling.

Random sampling occurs when, given a sample size *n*, all possible samples of that size are equally likely to be chosen. Random numbers from calculators are typically used in this setting. Each member of a population is paired with a number, and then a set of random numbers is generated. Each person paired with one of those random numbers is selected. A **systematic sample** is when every fourth, seventh, tenth, etc., person from a population is selected to be in a sample. A **stratified sample** is when the population is divided into subgroups, or **strata**, using a characteristic, and then members from each stratum are randomly selected. For instance, university students could be divided into age groups and then selected from each age group. Finally, a **cluster sample** is when a sample is used from an already selected group, like city block or zip code. These four methods are used most frequently because they are most likely to yield unbiased results.

Once an unbiased sample is obtained, data is collected. Common data collection methods include surveys with questions that are unbiased, contain clear language, avoid double negatives, and do not contain compound sentences that ask two questions at once. The simpler verbiage, the better when formulating these questions.

Interpreting Categorical and Quantitative Data

Interpreting Tables, Charts, and Graphs

Tables, charts, and graphs can be used to convey information about different variables. They are all used to organize, categorize, and compare data, and they all come in different shapes and sizes. Each type has its own way of showing information, whether it is in a column, shape, or picture. To answer a question relating to a table, chart, or graph, some steps should be followed. First, the problem should be read thoroughly to determine what is being asked to determine what quantity is unknown. Then, the title of the table, chart, or graph should be read. The title should clarify what actual data is being summarized in the table. Next, look at the key and both the horizontal and vertical axis labels, if they are given. These items will provide information about how the data is organized. Finally, look to see if there is any more labeling inside the table. Taking the time to get a good idea of what the table is summarizing will be helpful as it is used to interpret information.

Tables are a good way of showing a lot of information in a small space. The information in a table is organized in columns and rows. For example, a table may be used to show the number of votes each candidate received in an election. By interpreting the table, one may observe which candidate won the election and which candidates came in second and third. In using a bar chart to display monthly rainfall amounts in different countries, rainfall can be compared between countries at different times of the year. Graphs are also a useful way to show change in variables over time, as in a line graph, or percentages of a whole, as in a pie graph.

The table below relates the number of items to the total cost. The table shows that 1 item costs $5. By looking at the table further, 5 items cost $25, 10 items cost $50, and 50 items cost $250. This cost can be extended for any number of items. Since 1 item costs $5, then 2 items would cost $10. Though this information isn't in the table, the given price can be used to calculate unknown information.

Number of Items	1	5	10	50
Cost ($)	5	25	50	250

A **bar graph** is a graph that summarizes data using bars of different heights. It is useful when comparing two or more items or when seeing how a quantity changes over time. It has both a horizontal and vertical axis. Interpreting bar graphs includes recognizing what each bar represents and connecting that to the two variables. The bar graph below shows the scores for six people on three different games. The color of the bar shows which game each person played, and the height of the bar indicates their score for that game. William scored 25 on game 3, and Abigail scored 38 on game 3. By comparing the bars, it's obvious that Williams scored lower than Abigail.

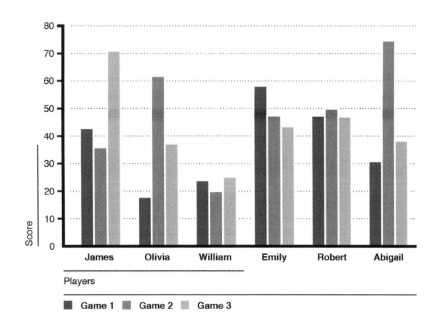

A **line graph** is a way to compare two variables. Each variable is plotted along an axis, and the graph contains both a horizontal and a vertical axis. On a line graph, the line indicates a continuous change. The change can be seen in how the line rises or falls, known as its slope, or rate of change. Often, in line graphs, the horizontal axis represents a variable of time. Readers can quickly see if an amount has grown or decreased over time. The bottom of the graph, or the x-axis, shows the units for time, such as days, hours, months, etc. If there are multiple lines, a comparison can be made between what the two lines represent. For example, the following line graph, shown previously, displays the change in temperature over five days. The top line represents the high, and the bottom line represents the low for each day. Looking at the top line alone, the high decreases for a day, then increases on Wednesday. Then it decreased on Thursday and increases again on Friday. The low temperatures have a similar trend, shown in bottom line. The range in temperatures each day can also be calculated by finding the difference between the top line and bottom line on a particular day. On Wednesday, the range was 14 degrees, from 62 to 76° F.

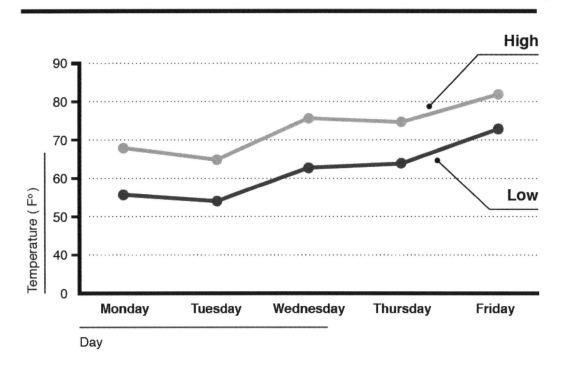

Daily Temperatures

Pie charts are used to show percentages of a whole, as each category is given a piece of the pie, and together all the pieces make up a whole. They are a circular representation of data that are used to highlight numerical proportions. It is true that the arc length of each pie slice is proportional to the amount it individually represents. When a pie chart is shown, a reader can quickly make comparisons by comparing the sizes of the pieces of the pie. They can be useful for comparison between different categories. The following pie chart is a simple example of three different categories shown in comparison to each other.

Light gray represents cats, dark gray represents dogs, and the gray between those two represents other pets. As the pie is cut into three equal pieces, each value represents just more than 33 percent, or $\frac{1}{3}$ of the whole. Values 1 and 2 may be combined to represent $\frac{2}{3}$ of the whole. In an example where the total pie represents 75,000 animals, then cats would be equal to $\frac{1}{3}$ of the total, or 25,000. Dogs would equal 25,000 and other pets would hold equal 25,000.

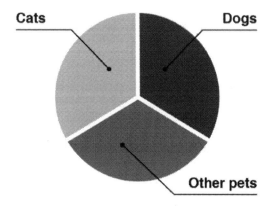

The fact that a circle is 360 degrees is used to create a pie chart. Because each piece of the pie is a percentage of a whole, that percentage is multiplied times 360 to get the number of degrees each piece represents. In the example above, each piece is 1/3 of the whole, so each piece is equivalent to 120 degrees. Together, all three pieces add up to 360 degrees.

Stacked bar graphs are also used fairly frequently when comparing multiple variables at one time. They combine some elements of both pie charts and bar graphs, using the organization of bar graphs and the proportionality aspect of pie charts. The following is an example of a stacked bar graph that represents the number of students in a band playing drums, flute, trombone, and clarinet. Each bar graph is broken up further into girls and boys.

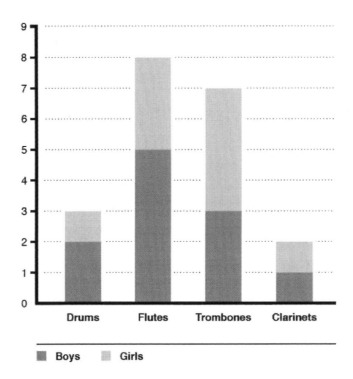

To determine how many boys play trombone, refer to the darker portion of the trombone bar, which indicates 3 boys.

A **scatterplot**, also known as a **scattergram**, is another way to represent paired data. It uses Cartesian coordinates, like a line graph, meaning it has both a horizontal and vertical axis. Each data point is represented as a dot on the graph. The dots are never connected with a line. For example, the following is a scatterplot showing people's age versus height.

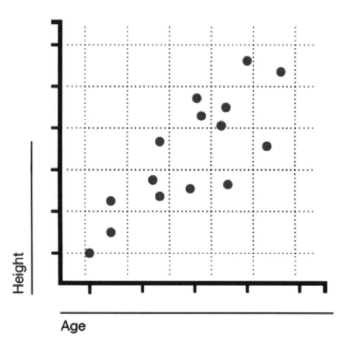

Scatterplots can be used to predict another value and to see if a correlation exists between a set of data. If the data resembles a straight line, the data is associated or correlated. The following is an example of a scatterplot in which the data does not seem to have an association:

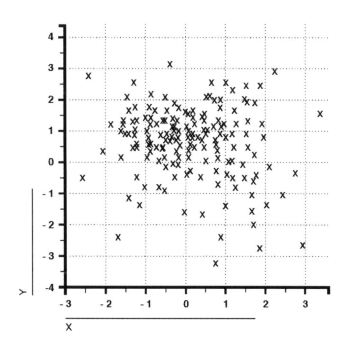

Sets of numbers and other similarly organized data can also be represented graphically. Venn diagrams are a common way to do so. A **Venn diagram** represents each set of data as a circle. The circles overlap, showing that each set of data is overlapping. A Venn diagram is also known as a **logic diagram** because it visualizes all possible logical combinations between two sets. Common elements of two sets are represented by the area of overlap. The following is an example of a Venn diagram of two sets A and B:

Parts of the Venn Diagram

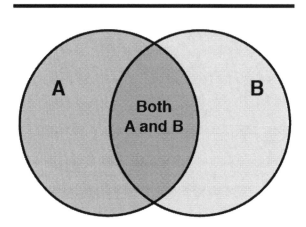

Another name for the area of overlap is the **intersection.** The intersection of A and B, written $A \cap B$, contains all elements that are in both sets A and B. The **union** of A and B, $A \cup B$, contains all elements that are in either set A or set B. Finally, the **complement** of $A \cup B$ is equal to all elements that are not in either set A or set B. These elements are placed outside of the circles.

The following is an example of a Venn diagram in which 22 students were surveyed asking about their siblings. Ten students only had a brother, 7 students only had a sister, and 5 had both a brother and a sister. This number 5 is the intersection and is placed where the circles overlap. Two students did not have a brother or a sister. Therefore, two is the complement and is placed outside of the circles.

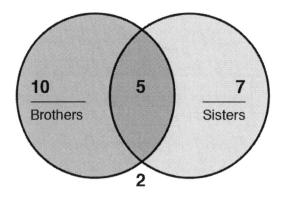

Venn diagrams can have more than two sets of data. The more circles, the more logical combinations are represented by the overlapping. The following is a Venn diagram that represents sock colors worn by a class of students. There were 30 students surveyed. The innermost region represents those students that had green, pink, and blue on their socks (perhaps in a striped pattern). Therefore, 2 students had all three colors. In this example, all students had at least one color on their socks, so no one exists in the complement.

30 students

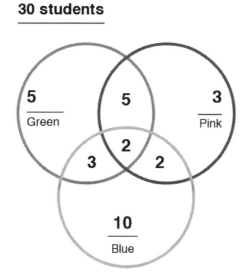

Venn diagrams are typically not drawn to scale, but if they are and their area is proportional to the amount of data it represents, it is known as an **area-proportional Venn diagram**.

Describing Distributions
One way information can be interpreted from tables, charts, and graphs is through statistics. The three most common calculations for a set of data are the mean, median, and mode. These three are called **measures of central tendency**. Measures of central tendency are helpful in comparing two or more different sets of data. The **mean** refers to the average and is found by adding up all values and dividing the total by the number of values. In other words, the mean is equal to the sum of all values divided by the number of data entries. For example, if you bowled a total of 532 points in 4 bowling games, your mean score was $\frac{532}{4} = 133$ points per game. A common application of mean useful to students is calculating what he or she needs to receive on a final exam to receive a desired grade in a class.

The **median** is found by lining up values from least to greatest and choosing the middle value. If there's an even number of values, then the mean of the two middle amounts must be calculated to find the median. For example, the median of the set of dollar amounts $5, $6, $9, $12, and $13 is $9. The median of the set of dollar amounts $1, $5, $6, $8, $9, $10 is $7, which is the mean of $6 and $8. The **mode** is the value that occurs the most. The mode of the data set {1, 3, 1, 5, 5, 8, 10} actually refers to two numbers: 1 and 5. In this case, the data set is **bimodal** because it has two modes. A data set can have no mode if no amount is repeated. Another useful statistic is range. The **range** for a set of data refers to the difference between the highest and lowest value.

In some cases, some numbers in a list of data might have weights attached to them. In that case, a **weighted mean** can be calculated. A common application of a weighted mean is GPA. In a semester, each class is assigned a number of credit hours, its weight, and at the end of the semester each student receives a grade. To compute GPA, an A is a 4, a B is a 3, a C is a 2, a D is a 1, and an F is a 0. Consider a student that takes a 4-hour English class, a 3-hour math class, and a 4-hour history class and receives all

B's. The weighted mean, GPA, is found by multiplying each grade times its weight, number of credit hours, and dividing by the total number of credit hours. Therefore, the student's GPA is:

$$\frac{3 \times 4 + 3 \times 3 + 3 \times 4}{11} = \frac{33}{1} = 3.0.$$

The following bar chart shows how many students attend a cycle class on each day of the week. To find the mean attendance for the week, each day's attendance can be added together, $10 + 7 + 6 + 9 + 8 + 14 + 4 = 58$, and the total divided by the number of days, $58 \div 7 = 8.3$. The mean attendance for the week was 8.3 people. The median attendance can be found by putting the attendance numbers in order from least to greatest: 4, 6, 7, 8, 9, 10, 14, and choosing the middle number: 8 people. The mode for attendance is none for this set of data because no numbers repeat. The range is 10, which is found by finding the difference between the lowest number, 4, and the highest number, 14.

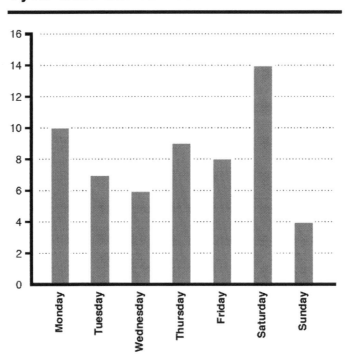

Cycle class attendance

A **histogram** is a bar graph used to group data into "bins" that cover a range on the horizontal, or x-axis. Histograms consist of rectangles whose height is equal to the frequency of a specific category. The horizontal axis represents the specific categories. Because they cover a range of data, these bins have no gaps between bars, unlike the bar graph above. In a histogram showing the heights of adult golden retrievers, the bottom axis would be groups of heights, and the y-axis would be the number of dogs in each range. Evaluating this histogram would show the height of most golden retrievers as falling within a certain range. It also provides information to find the average height and range for how tall golden retrievers may grow.

The following is a histogram that represents exam grades in a given class. The horizontal axis represents ranges of the number of points scored, and the vertical axis represents the number of students. For example, approximately 33 students scored in the 60 to 70 range.

Results of the exam

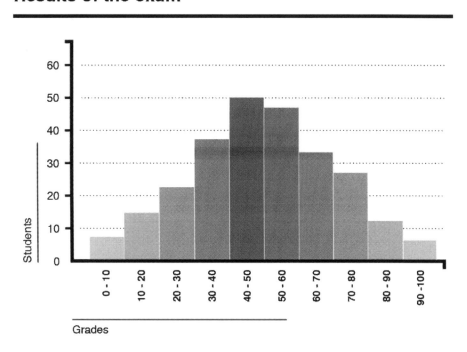

Measures of central tendency can be discussed using a histogram. If the points scored were shown with individual rectangles, the tallest rectangle would represent the mode. A bimodal set of data would have two peaks of equal height. Histograms can be classified as having data **skewed to the left, skewed to the right**, or **normally distributed**, which is also known as **bell-shaped**.

These three classifications can be seen in the following chart:

Measures of central tendency images

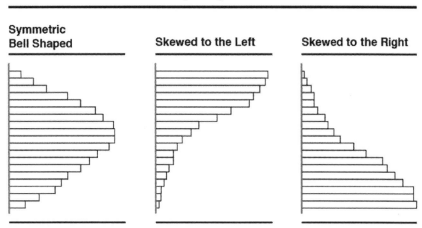

When the data is normal, the mean, median, and mode are all very close. They all represent the most typical value in the data set. The mean is typically used as the best measure of central tendency in this case because it does include all data points. However, if the data is skewed, the mean becomes less meaningful. The median is the best measure of central tendency because it is not affected by any outliers, unlike the mean. When the data is skewed, the mean is dragged in the direction of the skew. Therefore, if the data is not normal, it is best to use the median as the measure of central tendency.

The measures of central tendency and the range may also be found by evaluating information on a line graph.

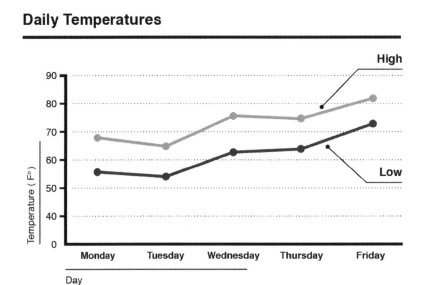

Daily Temperatures

In the line graph above that shows the daily high and low temperatures, the average high temperature can be found by gathering data from each day on the triangle line. The days' highs are 82, 78, 75, 65, and 70. The average is found by adding them together to get 370, then dividing by 5 (because there are 5 temperatures). The average high for the five days is 74. If 74 degrees is found on the graph, then it falls in the middle of the values on the triangle line. The low temperature can be found in the same way.

Given a set of data, the **correlation coefficient**, r, measures the association between all the data points. If two values are **correlated**, there is an association between them. However, correlation does not necessarily mean causation, or that that one value causes the other. There is a common mistake made that assumes correlation implies causation. Average daily temperature and number of sunbathers are both correlated and have causation. If the temperature increases, that change in weather causes more people to want to catch some rays. However, wearing plus-size clothing and having heart disease are two variables that are correlated but do not have causation. The larger someone is, the more likely he or she is to have heart disease. However, being overweight does not cause someone to have the disease.

The value of the correlation coefficient is between −1 and 1, where −1 represents a perfect negative linear relationship, 0 represents no relationship between the two data sets, and 1 represents a perfect positive linear relationship. A negative linear relationship means that as x-values increase, y-values decrease. A positive linear relationship means that as x-values increase, y-values increase. The formula for computing the correlation coefficient is:

$$r = \frac{n\sum xy - (\sum x)(\sum y)}{\sqrt{n(\sum x^2) - (\sum x)^2}\sqrt{n(\sum y^2) - (y)^2}}$$

where n is the number of data points. The closer r is to 1 or −1, the stronger the correlation. A correlation can be seen when plotting data. If the graph resembles a straight line, there is a correlation.

Solving Problems Involving Measures of Center and Range

As mentioned, a data set can be described by calculating the mean, median, and mode. These values allow the data to be described with a single value that is representative of the data set.

The most common measure of center is the **mean,** also referred to as the **average.**

To calculate the mean:

> 1. Add all data values together
>
> 2. Divide by the sample size (the number of data points in the set)

The **median** is middle data value, so that half of the data lies below this value and half lies below the data value.

To calculate the median:

> 1. Order the data from least to greatest
>
> 2. The point in the middle of the set is the median
>
> 3. If there is an even number of data points, add the two middle points and divide by 2

The **mode** is the data value that occurs most often.

To calculate the mode:

> 1. Order the data from least to greatest
>
> 2. Find the value that occurs most often

Example: Amelia is a leading scorer on the school's basketball team. The following data set represents the number of points that Amelia has scored in each game this season. Use the mean, median, and mode to describe the data.

16, 12, 26, 14, 13, 28, 14, 12, 15, 25

Solution:

Mean:

$$16 + 12 + 26 + 14 + 28 + 14 + 12 + 15 + 25 = 162$$

$$162 \div 9 = 18$$

Amelia averages 18 points per game.

Median:

12, 12, 14, 14, **15**, 16, 25, 26, 28

Amelia's median score is 15 points.

Mode:

12, 12, 14, 14, 15, 16, 25, 26, 28

The numbers 12 and 14 each occur twice, so this data set has 2 modes: 12 and 14 points.

The **range** is the difference between the largest and smallest values in the set. In the example above, the range is 28 – 12 = 16 points.

Determining How Changes in Data Affect Measures of Center or Range
An **outlier** is a data point that lies an unusual distance from other points in the data set. Removing an outlier from a data set will change the measures of center. Removing a large outlier (a high number) from a data set will decrease both the mean and the median. Removing a small outlier (a number much lower than most in the data set) from a data set will increase both the mean and the median. For example, given the data set {3, 6, 8, 12, 13, 14, 60}, the data point, 60, is an outlier because it is unusually far from the other points. In this data set, the mean is 16.6. Notice that this mean number is even larger than all other data points in the set except for 60. Removing the outlier, the mean changes to 9.3 and the median becomes 10. Removing an outlier will also decrease the range. In the data set above, the range is 57 when the outlier is included, but decreases to 11 when the outlier is removed.

Adding an outlier to a data set will affect the centers of measure as well. When a larger outlier is added to a data set, the mean and median increase. When a small outlier is added to a data set, the mean and median decrease. Adding an outlier to a data set will increase the range.

This does not seem to provide an appropriate measure of center when considering this data set. What will happen if that outlier is removed? Removing the extremely large data point, 60, is going to reduce the mean to 9.3. The mean decreased dramatically because 60 was much larger than any of the other data values. What would happen with an extremely low value in a data set like this one, {12, 87, 90, 95, 98, 100}? The mean of the given set is 80. When the outlier, 12, is removed, the mean should increase and should fit more closely to the other data points. Removing 12 and recalculating the mean shows that this is correct. The mean after removing 12 is 94. So, removing a large outlier will decrease the mean while removing a small outlier will increase the mean.

Patterns of Association in Bivariate Data

Variables are values that change, so determining the relationship between them requires an evaluation of who changes them. The **independent variable** is the variable controlled by the experimenter. It stands alone and isn't changed by other parts of the experiment. This variable is normally represented by x and is found on the horizontal, or x-axis, of a graph. If the variable changes because of a result in the experiment, then it's **dependent**. The dependent variable changes in response to the independent variable. It reacts to, or depends on, the independent variable. This variable is normally represented by y and is found on the vertical, or y-axis of the graph. As they interact, one is manipulated by the other. The manipulator is the independent, and the manipulated is the dependent. Once the independent and dependent variable are determined, they can be evaluated to have a positive, negative, or no correlation.

The relationship between two variables, x and y, can be seen on a scatterplot. The following scatterplot shows the relationship between weight and height. The graph shows the weight as x and the height as y. The first dot on the left represents a person who is 45 kg and approximately 150 cm tall. The other dots correspond in the same way. As the dots move to the right and weight increases, height also increases. A line could be drawn through the middle of the dots to move from bottom left to top right. This line would indicate a **positive correlation** between the variables. If the variables had a **negative correlation**, then the dots would move from the top left to the bottom right.

Height and Weight

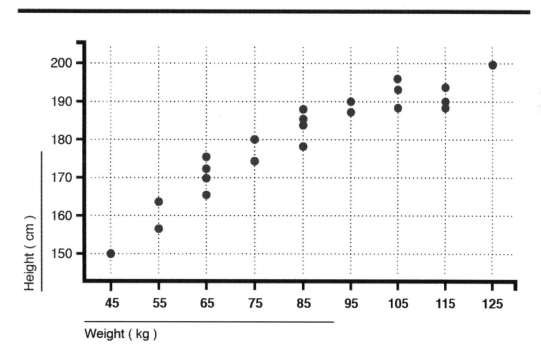

A **scatterplot** is useful in determining the relationship between two variables, but it's not required. Consider an example where a student scores a different grade on his math test for each week of the month. The independent variable would be the weeks of the month. The dependent variable would be the grades, because they change depending on the week. If the grades trended up as the weeks passed, then the relationship between grades and time would be positive. If the grades decreased as the time passed, then the relationship would be negative. (As the number of weeks went up, the grades went down.)

The relationship between two variables can further be described as strong or weak. The relationship between age and height shows a strong positive correlation because children grow taller as they grow up. In adulthood, the relationship between age and height becomes weak, and the dots will spread out. People stop growing in adulthood, and their final heights vary depending on factors like genetics and health. The closer the dots on the graph to the trend line, the stronger the relationship. As they spread apart, the relationship becomes weaker. If they are too spread out to determine a trend (and thus, correlation) up or down, then the variables are said to have no correlation.

Linear Regression Models

Graphs, equations, and tables are three different ways to represent linear relationships. The following graph shows a linear relationship because the relationship between the two variables is constant. As the distance increases by 25 miles, the time lapses by 1 hour. This pattern continues for the rest of the graph. The line represents a constant rate of 25 miles per hour. This graph can also be used to solve problems involving predictions for a future time. After 8 hours of travel, the rate can be used to predict the distance covered. Eight hours of travel at 25 miles per hour covers a distance of 200 miles. The equation at the top of the graph corresponds to this rate also. The same prediction of distance in a given time can be found using the equation. For a time of 10 hours, the distance would be 250 miles, as the equation yields $d = 25 \times 10 = 250 \ miles$.

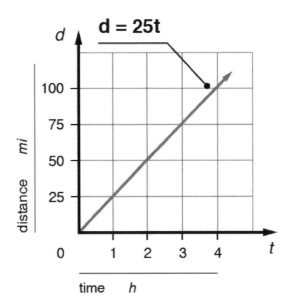

Another representation of a linear relationship can be seen in a table. The first thing to observe from the table is that the y-values increase by the same amount of 3 each time. As the x-values increase by 1, the y-values increase by 3. This pattern shows that the relationship is linear. If this table shows the money earned, y-value, for the hours worked, x-value, then it can be used to predict how much money will be earned for future hours. If 6 hours are worked, then the pay would be \$19. For further hours and money to be determined, it would be helpful to have an equation that models this table of values. The equation will show the relationship between x and y. The y-value can each time be determined by multiplying the x-

value by 3, then adding 1. The following equation models this relationship: $y = 3x + 1$. Now that there is an equation, any number of hours, x can be substituted into the equation to find the amount of money earned, y.

$y = 3x + 1$	
x	y
0	1
1	4
2	7
4	13
5	16

Calculating Probabilities

Probability describes how likely it is that an event will occur. Probabilities are always a number from 0 to 1. If an event has a high likelihood of occurrence, it will have a probability close to 1. If there is only a small chance that an event will occur, the likelihood is close to 0. A fair six-sided die has one of the numbers 1, 2, 3, 4, 5, and 6 on each side. When this die is rolled there is a one in six chance that it will land on 2. This is because there are six possibilities and only one side has a 2 on it. The probability then is $\frac{1}{6}$ or 0.167. The probability of rolling an even number from this die is three in six, or ½ or 0.5. This is because there are three sides on the die with even numbers (2, 4, 6), and there are six possible sides. The probability of rolling a number less than 10 is one because every side of the die has a number less than 6, so this is certain to occur. On the other hand, the probability of rolling a number larger than 20 is zero. There are no numbers greater than 20 on the die, so it is certain that this will not occur; thus, the probability is zero.

If a teacher says that the probability of anyone passing her final exam is 0.2, is it highly likely that anyone will pass? No, the probability of anyone passing her exam is low because 0.2 is closer to 0 than to 1. If another teacher is proud that the probability of students passing his class is 0.95, how likely is it that a student will pass? It is highly likely that a student will pass because the probability, 0.95, is very close to 1.

Using Two-Way Tables to Summarize Categorical Data and Relative Frequencies, and to Calculate Conditional Probability

A **two-way frequency table** displays categorical data with two variables, and it highlights relationships that exist between those two variables. Such tables are used frequently to summarize survey results, and are also known as **contingency tables**. Each cell shows a count pertaining to that individual variable paring, known as a **joint frequency**, and the totals of each row and column also are in the table. Consider the following two-way frequency table:

Distribution of the Residents of a Particular Village

	70 or older	69 or younger	Totals
Women	20	40	60
Men	5	35	40
Total	25	75	100

The table above shows the breakdown of ages and sexes of 100 people in a particular village. The total number of people in the data is shown in the bottom right corner. Each total is shown at the end of each row or column, as well. For instance, there were 25 people age 70 or older and 60 women in the data. The 20 in the first cell shows that out of 100 total villagers, 20 were women aged 70 or older. The 5 in the cell below shows that out of 100 total villagers, 5 were men aged 70 or older.

A two-way table can also show **relative frequencies**. If instead of the count, the percentage of people in each category was placed into the cells, the two-way table would show relative frequencies. If each frequency is calculated over the entire total of 100, the first cell would be 20% or 0.2. However, the relative frequencies can also be calculated over row or column totals. If row totals were used, the first cell would be $\frac{20}{60} = 0.333 = 33.3\%$. If column totals were used, the first cell would be $\frac{20}{25} = 0.8 = 80\%$.

Such tables can be used to calculate **conditional probabilities**, which are probabilities that an event occurs, given another event. Consider a randomly selected villager. The probability of selecting a male 70 years old or older is $\frac{5}{100} = 0.05$ because there are 5 males over the age of 70 and 100 total villagers.

Section 1 Practice Questions

1. What is $\frac{12}{60}$ converted to a percentage?
   - a. 0.20
   - b. 20%
   - c. 25%
   - d. 12%
   - e. 1.2%

2. Which of the following is the correct decimal form of the fraction $\frac{14}{33}$ rounded to the nearest hundredth place?
   - a. 0.420
   - b. 0.14
   - c. 0.424
   - d. 0.140
   - e. 0.42

3. Which of the following represents the correct sum of $\frac{14}{15}$ and $\frac{2}{5}$?
   - a. $\frac{20}{15}$
   - b. $\frac{4}{3}$
   - c. $\frac{16}{20}$
   - d. $\frac{4}{5}$
   - e. $\frac{16}{15}$

4. What is the product of $\frac{5}{14}$ and $\frac{7}{20}$?
   - a. $\frac{1}{8}$
   - b. $\frac{35}{280}$
   - c. $\frac{12}{34}$
   - d. $\frac{1}{2}$
   - e. $\frac{7}{140}$

5. What is the area of the following figure?

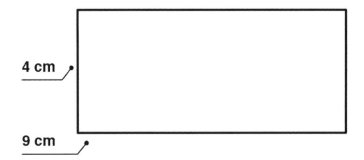

4 cm

9 cm

    a. 26 cm
    b. 36 cm
    c. 13 cm^2
    d. 36 cm^2
    e. 65 cm^2

6. What is the volume of the given figure?

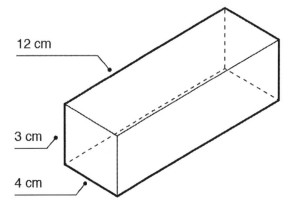

12 cm

3 cm

4 cm

    a. 36 cm^2
    b. 144 cm^3
    c. 72 cm^3
    d. 36 cm^3
    e. 144 cm^2

7. A study of adult drivers finds that it is likely that an adult driver wears his seatbelt. Which of the following could be the probability that an adult driver wears his seat belt?
    a. 0.90
    b. 0.05
    c. 0.25
    d. 0
    e. 1.5

8. Which of the following is incorrect?

    a. $-\dfrac{1}{5} < \dfrac{4}{5}$

    b. $\dfrac{4}{5} > -\dfrac{1}{5}$

    c. $-\dfrac{1}{5} > \dfrac{4}{5}$

    d. $\dfrac{1}{5} > -\dfrac{4}{5}$

    e. $\dfrac{4}{5} > \dfrac{1}{5}$

9. What is the solution to the equation $3(x + 2) = 14x - 5$?

    a. $x = 1$
    b. $x = -1$
    c. $x = 0$
    d. All real numbers
    e. No solution

10. What is the solution to the equation $10 - 5x + 2 = 7x + 12 - 12x$?

    a. $x = 12$
    b. $x = 1$
    c. $x = 0$
    d. All real numbers
    e. No solution

11. Which of the statements below is a statistical question?

    a. What was your grade on the last test?
    b. What were the grades of the students in your class on the last test?
    c. What kind of car do you drive?
    d. What was Sam's time in the marathon?
    e. What textbooks does Marty use this semester?

12. How many cases of cola can Lexi purchase if each case is $3.50 and she has $40?

    a. 10
    b. 12
    c. 11.4
    d. 11
    e. 12.5

13. Two consecutive integers exist such that the sum of three times the first and two less than the second is equal to 411. What are those integers?

    a. 103 and 104
    b. 104 and 105
    c. 102 and 103
    d. 100 and 101
    e. 101 and 102

14. In a neighborhood, 15 out of 80 of the households have children under the age of 18. What percentage of the households have children?

    a. 0.1875%

    b. 18.75%

    c. 1.875%

    d. 15%

    e. 1.50%

15. The number of members of the House of Representatives varies directly with the total population in a state. If the state of New York has 19,800,000 residents and has 27 total representatives, how many should Ohio have with a population of 11,800,000?

    a. 10

    b. 9

    c. 11

    d. 5

    e. 12

16. Paul took a written driving test, and he got 12 of the questions correct. If he answered 75% of the total questions correctly, how many problems were there in the test?

    a. 25

    b. 15

    c. 20

    d. 18

    e. 16

17. If a car is purchased for $15,395 with a 7.25% sales tax, how much is the total price?

    a. $15,395.07

    b. $16,511.14

    c. $16,411.13

    d. $15,402

    e. $16,113.10

18. What is the solution to the following system of equations?
$$2x - y = 6$$
$$y = 8x$$

    a. (1, 8)

    b. (-1, -8)

    c. (-1, 8)

    d. All real numbers.

    e. There is no solution.

19. The following set represents the test scores from a university class: {35, 79, 80, 87, 87, 90, 92, 95, 95, 98, 99}. If the outlier is removed from this set, which of the following is TRUE?

    a. The mean and the median will decrease.

    b. The mean and the median will increase.

    c. The mean and the mode will increase.

    d. The mean and the mode will decrease.

    e. The mean, median, and mode will increase.

20. Bindee is having a barbeque on Sunday and needs 12 packets of ketchup for every 5 guests. If 60 guests are coming, how many packets of ketchup should she buy?

    a. 100
    b. 12
    c. 144
    d. 60
    e. 300

21. What is the perimeter of the following figure?

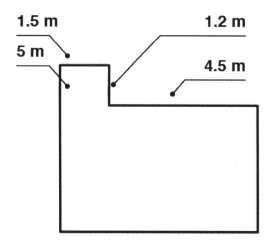

    a. 13.4 m
    b. 22 m
    c. 12.2 m
    d. 22.5 m
    e. 24.4 m

22. If Oscar's bank account totaled $4,000 in March and $4,900 in June, what was the rate of change in his bank account total over those three months?

    a. $900 a month
    b. $300 a month
    c. $4,900 a month
    d. $100 a month
    e. $4,000 a month

23. Which of the following is the equation of a vertical line that runs through the point (1, 4)?

    a. $x = 1$
    b. $y = 1$
    c. $x = 4$
    d. $y = 4$
    e. $x = y$

24. From the chart below, which two are preferred by more men than women?

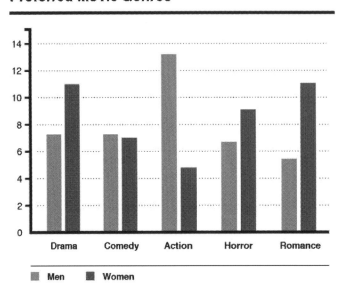

Preferred Movie Genres

Men   Women

    a. Comedy and Action
    b. Drama and Comedy
    c. Action and Horror
    d. Action and Romance
    e. Romance and Comedy

25. Which type of graph best represents a continuous change over a period of time?
    a. Stacked bar graph
    b. Bar graph
    c. Pie graph
    d. Histogram
    e. Line graph

Section 1 Answer Explanations

1. B: The fraction $\frac{12}{60}$ can be reduced to $\frac{1}{5}$, in lowest terms. First, it must be converted to a decimal. Dividing 1 by 5 results in 0.2. Then, to convert to a percentage, move the decimal point two units to the right and add the percentage symbol. The result is 20%.

2. E: If a calculator were to be used, you would divide 33 into 14 and keep two decimal places. However, since a calculator is not used on the test, multiply both the numerator and denominator times 3. This results in the fraction $\frac{42}{99}$, and hence a decimal of 0.42.

3. B: Common denominators must be used. The LCD is 15, and $\frac{2}{5} = \frac{6}{15}$. Therefore, $\frac{14}{15} + \frac{6}{15} = \frac{20}{15}$, and in lowest terms the answer is $\frac{4}{3}$. A common factor of 5 was divided out of both the numerator and denominator.

4. A: A product is found by multiplication. Multiplying two fractions together is easier when common factors are cancelled first to avoid working with larger numbers.

$$\frac{5}{14} \times \frac{7}{20} = \frac{5}{2 \times 7} \times \frac{7}{5 \times 4}$$

$$\frac{1}{2} \times \frac{1}{4} = \frac{1}{8}$$

5. D: The area for a rectangle is found by multiplying the length by the width. The area is also measured in square units, so the correct answer is Choice *D*. The answer of 26 is the perimeter. The answer of 13 is found by adding the two dimensions instead of multiplying.

6. B: The volume of a rectangular prism is found by multiplying the length by the width by the height. This formula yields an answer of 144 cubic units. The answer must be in cubic units because volume involves all three dimensions. Each of the other answers have only two dimensions that are multiplied, and one dimension is forgotten, as in *D*, where 12 and 3 are multiplied, or have incorrect units, as in *E*.

7. A: The probability of .9 is closer to 1 than any of the other answers. The closer a probability is to 1, the greater the likelihood that the event will occur. The probability of 0.05 shows that it is very unlikely that an adult driver will wear their seatbelt because it is close to zero. A zero probability means that it will not occur. The probability of 0.25 is closer to zero than to one, so it shows that it is unlikely an adult will wear their seatbelt. Choice *E* is wrong because probability must fall between 0 and 1.

8. C: $-\frac{1}{5} > \frac{4}{5}$ is incorrect. The expression on the left is negative, which means that it is smaller than the expression on the right. As it is written, the inequality states that the expression on the left is greater than the expression on the right, which is not true.

9. A: First, the distributive property must be used on the left side. This results in:

$$3x + 6 = 14x - 5$$

The addition property is then used to add 5 to both sides, and then to subtract $3x$ from both sides, resulting in $11 = 11x$. Finally, the multiplication property is used to divide each side by 11. Therefore, $x = 1$ is the solution.

10. D: First, like terms are collected to obtain:

$$12 - 5x = -5x + 12$$

Then, if the addition principle is used to move the terms with the variable, $5x$ is added to both sides and the mathematical statement $12 = 12$ is obtained. This is always true; therefore, all real numbers satisfy the original equation.

11. B: This is a statistical question because to determine this answer one would need to collect data from each person in the class and it is expected the answers would vary. The other answers do not require data to be collected from multiple sources, therefore the answers will not vary.

12. D: This is a one-step real-world application problem. The unknown quantity is the number of cases of cola to be purchased. Let x be equal to this amount. Because each case costs $3.50, the total number of cases times $3.50 must equal $40. This translates to the mathematical equation $3.5x = 40$. Divide both sides by 3.5 to obtain $x = 11.4286$, which has been rounded to four decimal places. Because cases are sold whole (the store does not sell portions of cases), and there is not enough money to purchase 12 cases, there is only enough money to purchase 11.

13. A: First, the variables have to be defined. Let x be the first integer and therefore $x + 1$ is the second integer. This is a two-step problem. The sum of three times the first and two less than the second is translated into the following expression:

$$3x + (x + 1 - 2)$$

This expression is set equal to 411 to obtain:

$$3x + (x + 1 - 2) = 412$$

The left-hand side is simplified to obtain $4x - 1 = 411$. The addition and multiplication properties are used to solve for x. First, add 1 to both sides and then divide both sides by 4 to obtain $x = 103$. The next consecutive integer is 104.

14. B: First, the information is translated into the ratio $\frac{15}{80}$. To find the percentage, translate this fraction into a decimal by dividing 15 by 80. The corresponding decimal is 0.1875. Move the decimal point two units to the right to obtain the percentage 18.75%.

15. B: The number of representatives varies directly with the population, so the equation necessary is $N = k \cdot P$, where N is number of representatives, k is the variation constant, and P is total population in millions. Plugging in the information for New York allows k to be solved for. This process gives $19.8 = k \times 27$, so $k = 0.73$. Therefore, the formula for number of representatives given total population in billions is $N = .73 \times P$. Plugging in $P = 11.6$ for Ohio results in $N = 8.6$, which rounds up to 9 total Representatives.

16. E: The unknown quantity is the number of total questions on the test. Let x be equal to this unknown quantity. Therefore, $0.75x = 12$. Divide both sides by 0.75 to obtain $x = 16$.

17. B: If sales tax is 7.25%, the price of the car must be multiplied times 1.0725 to account for the additional sales tax. Therefore:

$$15,395 \times 1.0725 = 16,511.1375$$

This amount is rounded to the nearest cent, which is $16,511.14.

18. B: This system can be solved using substitution. Plug the second equation in for y in the first equation to obtain $2x - 8x = 6$, which simplifies to $-6x = 6$. Divide both sides by 6 to get $x = -1$, which is then back-substituted into either original equation to obtain $y = -8$.

19. B: The outlier is 35. When a small outlier is removed from a data set, the mean and the median increase. The first step in this process is to identify the outlier, which is the number that lies away from the given set. Once the outlier is identified, the mean and median can be recalculated. The mean will be affected because it averages all of the numbers. The median will be affected because it finds the middle number, which is subject to change because a number is lost. The mode will most likely not change because it is the number that occurs the most, which will not be the outlier if there is only one outlier.

20. C: This problem involves ratios and percentages. If 12 packets are needed for every 5 people, this statement is equivalent to the ratio $\frac{12}{5}$. The unknown amount x is the number of ketchup packets needed for 60 people. The proportion $\frac{12}{5} = \frac{x}{60}$ must be solved. Cross-multiply to obtain $12 \times 60 = 5x$. Therefore, $720 = 5x$. Divide each side by 5 to obtain $x = 144$ packets.

21. B: The perimeter is found by adding the length of all the exterior sides. When the given dimensions are added, the perimeter is 22 meters. The equation to find the perimeter can be:

$$P = 5 + 1.5 + 1.2 + 4.5 + 3.8 + 6 = 22$$

The last two dimensions can be found by subtracting 1.2 from 5, and adding 1.5 and 4.5, respectively.

22. B: The average rate of change is found by calculating the difference in dollars over the elapsed time. Therefore, the rate of change is equal to $4,900-$4,000÷3 months, which is equal to $900÷3 or $300 a month.

23. A: A vertical line has the same x value for any point on the line. Other points on the line would be (1, 3), (1, 5), (1, 9,) etc. Mathematically, this is written as $x = 1$. A vertical line is always of the form $x = a$ for some constant a.

24. A: The chart is a bar chart showing how many men and women prefer each genre of movies. The dark gray bars represent the number of women, while the light gray bars represent the number of men. The light gray bars are higher and represent more men than women for the genres of Comedy and Action.

25. E: A line graph represents continuous change over time. The line on the graph is continuous and not broken, as on a scatter plot. Stacked bar graphs are used when comparing multiple variables at one time. They combine some elements of both pie charts and bar graphs, using the organization of bar graphs and the proportionality aspect of pie charts. A bar graph may show change but isn't necessarily continuous over time. A pie graph is better for representing percentages of a whole. Histograms are best used in grouping sets of data in bins to show the frequency of a certain variable.

Section 2 Practice Questions

1. What is the result of dividing 24 by $\frac{8}{5}$?

    a. $\frac{5}{3}$

    b. $\frac{3}{5}$

    c. $\frac{120}{8}$

    d. 15

    e. $\frac{24}{5}$

2. Subtract $\frac{5}{14}$ from $\frac{5}{24}$. Which of the following is the correct result?

    a. $\frac{25}{168}$

    b. 0

    c. $-\frac{25}{168}$

    d. $\frac{1}{10}$

    e. $-\frac{1}{10}$

3. Which of the following is a correct mathematical statement?

    a. $\frac{1}{3} < -\frac{4}{3}$

    b. $-\frac{1}{3} > \frac{4}{3}$

    c. $\frac{1}{3} > \frac{4}{3}$

    d. $-\frac{1}{3} \geq \frac{4}{3}$

    e. $\frac{1}{3} > -\frac{4}{3}$

4. Gina took an algebra test last Friday. There were 35 questions, and she answered 60% of them correctly. How many correct answers did she have?

    a. 35

    b. 20

    c. 21

    d. 25

    e. 18

5. Which of the following is the result when solving the equation $4(x + 5) + 6 = 2(2x + 3)$?

    a. $x = 6$

    b. $x = 1$

    c. $x = 26$

    d. All real numbers

    e. No solution

6. A car manufacturer usually makes 15,412 SUVs, 25,815 station wagons, 50,412 sedans, 8,123 trucks, and 18,312 hybrids a month. About how many cars are manufactured each month?

    a. 120,000

    b. 200,000

    c. 300,000

    d. 12,000

    e. 20,000

7. Each year, a family goes to the grocery store every week and spends $105. About how much does the family spend annually on groceries?

    a. $10,000

    b. $50,000

    c. $500

    d. $5,000

    e. $1,200

8. Erin and Katie work at the same ice cream shop. Together, they always work less than 21 hours a week. In a week, if Katie worked two times as many hours as Erin, how many hours could Erin work?

    a. Less than 7 hours

    b. Less than or equal to 7 hours

    c. More than 7 hours

    d. Less than 8 hours

    e. More than 8 hours

9. Using the graph below, what is the mean number of visitors for the first 4 hours?

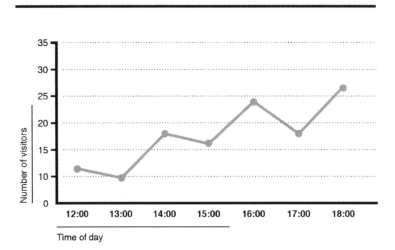

Museum Visitors

a. 12
b. 13
c. 14
d. 15
e. 16

10. What type of relationship is there between age and attention span as represented in the graph below?

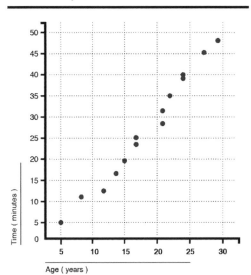

Attention Span

a. No correlation
b. Positive correlation
c. Negative correlation
d. Weak correlation
e. Inverse correlation

11. What is the area of the shaded region?

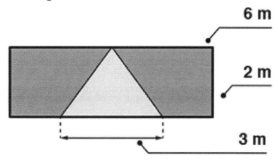

6 m

2 m

3 m

    a. 9 m²
    b. 12 m²
    c. 6 m²
    d. 8 m²
    e. 4.5 m²

12. What is the volume of the cylinder below?

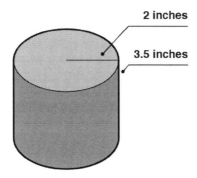

2 inches

3.5 inches

    a. 18.84 in³
    b. 45.00 in³
    c. 70.43 in³
    d. 43.96 in³
    e. 21.98 in³

13. Ms. Katy is wrapping end-of-the-year gifts for her class. There are 27 students in the class and each gift needs 2 feet, 4 inches of ribbon to tie around the package. What is the total length of ribbon that she needs to wrap all the gifts?
    a. 63 inches
    b. 63 feet
    c. 54 inches
    d. 65 feet
    e. 54 feet

14. How many centimeters are in 3 feet? (Note: 2.54cm = 1 inch)
    a. 0.635
    b. 1.1811
    c. 14.17
    d. 7.62
    e. 91.44

15. What is the correct factorization of the following binomial?
$$2y^3 - 128$$

    a. $2(y+8)(y-8)$
    b. $2(y-4)(y^2 + 4y + 16)$
    c. $2(y+4)(y-4)^2$
    d. $2(y-4)^3$
    e. $2(y-4)(y^2 + 4y + 16)$

16. What is the solution to the following linear inequality?
$$7 - \frac{4}{5}x < \frac{3}{5}$$

    a. $(-\infty, 8)$
    b. $(8, \infty)$
    c. $[8, \infty)$
    d. $(-\infty, 8]$
    e. $(-\infty, \infty)$

17. What is the solution to the following system of linear equations?
$$2x + y = 14$$
$$4x + 2y = -28$$

    a. (0, 0)
    b. (14, -28)
    c. (-14, 28)
    d. All real numbers
    e. There is no solution

18. Triple the difference of five and a number is equal to the sum of that number and 5. What is the number?
    a. 5
    b. 2
    c. 5.5
    d. 2.5
    e. 1

19. Which of the following is perpendicular to the line $4x + 7y = 23$?
    a. $y = -\frac{4}{7}x + 23$
    b. $y = \frac{7}{4}x - 12$
    c. $4x + 7y = 14$
    d. $y = -\frac{7}{4}x + 11$
    e. $y = \frac{4}{7}x - 12$

20. The mass of the moon is about 7.348×10^{22} kilograms and the mass of Earth is 5.972×10^{24} kilograms. How many times GREATER is Earth's mass than the moon's mass?

    a. 8.127×10^1

    b. 8.127

    c. 812.7

    d. 8.127×10^{-1}

    e. 0.8127

21. A grocery store sold 48 bags of apples in one day, and 9 of the bags contained Granny Smith apples. The rest contained Red Delicious apples. What is the ratio of bags of Granny Smith to bags of Red Delicious that were sold?

    a. 48:9

    b. 39:9

    c. 9:48

    d. 9:39

    e. 39:48

22. The percentage of smokers above the age of 18 in 2000 was 23.2 percent. The percentage of smokers above the age of 18 in 2015 was 15.1 percent. Find the average rate of change in the percent of smokers above the age of 18 from 2000 to 2015.

    a. -.54 percent

    b. -54 percent

    c. -5.4 percent

    d. .54 percent

    e. 5.4 percent

23. In order to estimate deer population in a forest, biologists obtained a sample of deer in that forest and tagged each one of them. The sample had 300 deer in total. They returned a week later and harmlessly captured 400 deer, and 5 were tagged. Use this information to estimate how many total deer were in the forest.

    a. 24,000 deer

    b. 30,000 deer

    c. 40,000 deer

    d. 100,000 deer

    e. 120,000 deer

24. What is the missing length *x*?

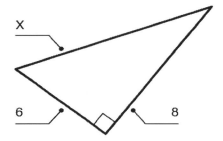

a. 6
b. 12
c. 10
d. 100
e. 14

25. Use the graph below entitled "Projected Temperatures for Tomorrow's Winter Storm" to answer the question.

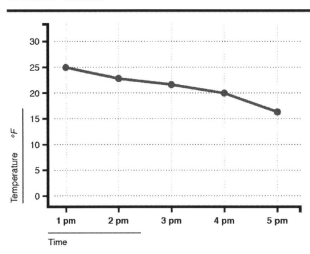

What is the expected temperature at 3:00 p.m.?
a. 25 degrees
b. 22 degrees
c. 20 degrees
d. 16 degrees
e. 18 degrees

Section 2 Answer Explanations

1. D: Division is completed by multiplying times the reciprocal. Therefore:

$$24 \div \frac{8}{5} = \frac{24}{1} \times \frac{5}{8}$$

$$\frac{3 \times 8}{1} \times \frac{5}{8} = \frac{15}{1} = 15$$

2. C: Common denominators must be used. The LCD is 168, so each fraction must be converted to have 168 as the denominator.

$$\frac{5}{24} - \frac{5}{14} = \frac{5}{24} \times \frac{7}{7} - \frac{5}{14} \times \frac{12}{12}$$

$$\frac{35}{168} - \frac{60}{168} = -\frac{25}{168}$$

3. E: The correct mathematical statement is the one in which the number to the left on the number line is less than the number to the right on the number line. It is written in answer *E* that $\frac{1}{3} > -\frac{4}{3}$, which is the same as $-\frac{4}{3} < \frac{1}{3}$, a correct statement.

4. C: Gina answered 60% of 35 questions correctly; 60% can be expressed as the decimal 0.60. Therefore, she answered $0.60 \times 35 = 21$ questions correctly.

5. E: The distributive property is used on both sides to obtain $4x + 20 + 6 = 4x + 6$. Then, like terms are collected on the left, resulting in $4x + 26 = 4x + 6$. Next, the addition principle is used to subtract $4x$ from both sides, and this results in the false statement $26 = 6$. Therefore, there is no solution.

6. A: Rounding can be used to find the best approximation. All of the values can be rounded to the nearest thousand. 15,412 SUVs can be rounded to 15,000. 25,815 station wagons can be rounded to 26,000. 50,412 sedans can be rounded to 50,000. 8,123 trucks can be rounded to 8,000. Finally, 18,312 hybrids can be rounded to 18,000. The sum of the rounded values is 117,000, which is closest to 120,000.

7. D: There are 52 weeks in a year, and if the family spends $105 each week, that amount is close to $100. A good approximation is $100 a week for 50 weeks, which is found through the product $50 \times 100 = $5,000$.

8. A: Let x be the unknown, the number of hours Erin can work. We know Katie works $2x$, and the sum of all hours is less than 21. Therefore, $x + 2x < 21$, which simplifies into $3x < 21$. Solving this results in the inequality $x < 7$ after dividing both sides by 3. Therefore, Erin can work less than 7 hours.

9. C: The mean for the number of visitors during the first 4 hours is 14. The mean is found by calculating the average for the four hours. Adding up the total number of visitors during those hours gives $12 + 10 + 18 + 16 = 56$. Then $56 \div 4 = 14$.

10. B: The relationship between age and time for attention span is a positive correlation because the general trend for the data is up and to the right. As the age increases, so does attention span.

11. A: The area of the shaded region is calculated in a few steps. First, the area of the rectangle is found using the formula:

$$A = length \times width = 6 \times 2 = 12$$

Second, the area of the triangle is found using the formula:

$$A = \frac{1}{2} \times base \times height = \frac{1}{2} \times 3 \times 2 = 3$$

The last step is to take the rectangle area and subtract the triangle area. The area of the shaded region is:

$$A = 12 - 3 = 9 \ m^2$$

12. D: The volume for a cylinder is found by using the formula:

$$V = \pi r^2 h = \pi(2^2) \times 3.5 = 43.96 \ in^3$$

13. B: Each package needs 2 feet, 4 inches, which is equal to $2\frac{1}{3} = \frac{7}{3}$ feet of ribbon. Because there are 27 students in the class, Ms. Kay needs $27 \times \frac{7}{3} = 63$ feet of ribbon to wrap all of the gifts.

14. E: The conversion between feet and centimeters requires a middle term. As there are 2.54 centimeters in 1 inch, the conversion between inches and feet must be found. As there are 12 inches in a foot, the fractions can be set up as follows:

$$3 \ feet \times \frac{12 \ inches}{1 \ foot} \times \frac{2.54 \ cm}{1 \ inch}$$

The feet and inches cancel out to leave only centimeters for the answer. The numbers are calculated across the top and bottom to yield:

$$\frac{3 \times 12 \times 2.54}{1 \times 1} = 91.44$$

The number and units used together form the answer of 91.44 cm.

15. E: First, the common factor 2 can be factored out of both terms, resulting in $2(y^3 - 64)$. The resulting binomial is a difference of cubes that can be factored using the rule $a^3 - b^3 = (a - b)(a^2 + ab + b^2)$ with a = y and b = 4. Therefore, the result is:

$$2(y - 4)(y^2 + 4y + 16)$$

16. B: The goal is to first isolate the variable. The fractions can easily be cleared by multiplying the entire inequality by 5, resulting in $35 - 4x < 3$. Then, subtract 35 from both sides and divide by -4. This results in $x > 8$. Notice the inequality symbol has been flipped because both sides were divided by a negative number. The solution set, all real numbers greater than 8, is written in interval notation as $(8, \infty)$. A parenthesis shows that 8 is not included in the solution set.

17. E: This system can be solved using the method of substitution. Solving the first equation for y results in $y = 14 - 2x$. Plugging this into the second equation gives $4x + 2(14 - 2x) = -28$, which simplifies to $28 = -28$, an untrue statement. Therefore, this system has no solution because no x-value will satisfy the system.

18. D: Let x be the missing quantity. The problem can be expressed as the following equation: $3(5 - x) = x + 5$. Distributing the 3 results in:

$$15 - 3x = x + 5$$

Subtract 5 from both sides, add $3x$ to both sides, and then divide both sides by 4. This results in:

$$\frac{10}{4} = \frac{5}{2} = 2.5$$

19. B: The slopes of perpendicular lines are negative reciprocals, meaning their product is equal to -1. The slope of the line given needs to be found. Its equivalent form in slope-intercept form is $y = -\frac{4}{7}x + 23$, so its slope is $-\frac{4}{7}$. The negative reciprocal of this number is $\frac{7}{4}$. The only line in the options given with this same slope is:

$$y = \frac{7}{4}x - 12$$

20. A: Division can be used to solve this problem. The division necessary is:

$$\frac{5.972 \times 10^{24}}{7.348 \times 10^{22}}$$

To compute this division, divide the constants first then use algebraic laws of exponents to divide the exponential expression. This results in about 0.8127×10^2, which written in scientific notation is 8.127×10^1.

21. D: There were 48 total bags of apples sold. If 9 bags were Granny Smith and the rest were Red Delicious, then $48 - 9 = 39$ bags were Red Delicious. Therefore, the ratio of Granny Smith to Red Delicious is 9:39.

22. A: The formula for the rate of change is the same as slope: change in y over change in x. The y-value in this case is percentage of smokers and the x-value is year. The change in percentage of smokers from 2000 to 2015 was 8.1 percent. The change in x was 2000-2015 = -15. Therefore:

$$8.1\%/_{-15} = -0.54\%$$

The percentage of smokers decreased 0.54 percent each year.

23. A: A proportion should be used to solve this problem. The ratio of tagged to total deer in each instance is set equal, and the unknown quantity is a variable x. The proportion is $\frac{300}{x} = \frac{5}{400}$. Cross-multiplying gives $120,000 = 5x$, and dividing through by 5 results in 24,000.

24. C: The Pythagorean Theorem can be used to find the missing length x because it is a right triangle. The theorem states that $6^2 + 8^2 = x^2$, which simplifies into $100 = x^2$. Taking the positive square root of both sides results in the missing value $x = 10$.

25. B: Look on the horizontal axis to find 3:00 p.m. Move up from 3:00 p.m. to reach the dot on the graph. Move horizontally to the left to the horizontal axis to between 20 and 25; the best answer choice is 22. The answer of 25 is too high above the projected time on the graph, and the answers of 20 and 16 degrees are too low.

Reading

Literary Fiction

Events, Plots, Characters, Settings, and Ideas

Putting Events in Order

Sequence structure is the order of events in which a story or information is presented to the audience. Sometimes the text will be presented in chronological order, or sometimes it will be presented by displaying the most recent information first, then moving backwards in time. The sequence structure depends on the author, the context, and the audience. The structure of a text also depends on the genre in which the text is written. Is it literary fiction? Is it a magazine article? Is it instructions for how to complete a certain task? Different genres will have different purposes for switching up the sequence of their writing.

The structure presented in literary fiction is also known as **narrative structure**. Narrative structure is the foundation on which the text moves. The basic ways for moving the text along are in the plot and the setting. The plot is the sequence of events in the narrative that move the text forward through cause and effect. The setting of a story is the place or time period in which the story takes place. Narrative structure has two main categories: linear and nonlinear.

Linear Narrative

Linear narrative is a narrative told in chronological order. Traditional linear narratives will follow the plot diagram below depicting the narrative arc. The narrative arc consists of the exposition, conflict, rising action, climax, falling action, and resolution.

- **Exposition**: The exposition is in the beginning of a narrative and introduces the characters, setting, and background information of the story. The importance of the exposition lies in its framing of the upcoming narrative. Exposition literally means "a showing forth" in Latin.

- **Conflict**: The conflict, in a traditional narrative, is presented toward the beginning of the story after the audience becomes familiar with the characters and setting. The conflict is a single instance between characters, nature, or the self, in which the central character is forced to make a decision or move forward with some kind of action. The conflict presents something for the main character, or protagonist, to overcome.

- **Rising Action**: The rising action is the part of the story that leads into the climax. The rising action will feature the development of characters and plot, and creates the tension and suspense that eventually lead to the climax.

- **Climax**: The climax is the part of the story where the tension produced in the rising action comes to a culmination. The climax is the peak of the story. In a traditional structure, everything before the climax builds up to it, and everything after the climax falls from it. It is the height of the narrative, and is usually either the most exciting part of the story or is marked by some turning point in the character's journey.

- **Falling Action**: The falling action happens as a result of the climax. Characters continue to develop, although there is a wrapping up of loose ends here. The falling action leads to the resolution.

- **Resolution**: The resolution is where the story comes to an end and usually leaves the reader with the satisfaction of knowing what happened within the story and why. However, stories do not always end in this fashion. Sometimes readers can be confused or frustrated at the end from lack of information or the absence of a happy ending.

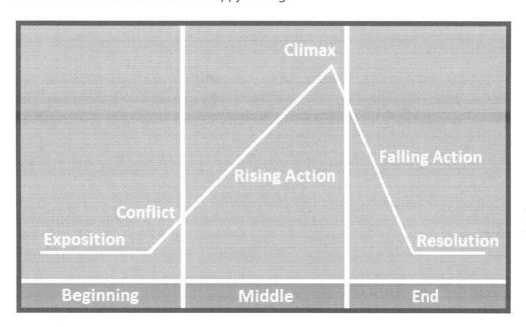

Nonlinear Narrative

A nonlinear narrative deviates from the traditional narrative in that it does not always follow the traditional plot structure of the narrative arc. Nonlinear narratives may include structures that are disjointed, circular, or disruptive, in the sense that they do not follow chronological order, but rather a nontraditional order of structure. *In medias res* is an example of a structure that predates the linear narrative. *In medias res* is Latin for "in the middle of things," which is how many ancient texts, especially epic poems, began their story, such as Homer's *Iliad*. Instead of having a clear exposition with a full development of characters, they would begin right in the middle of the action.

Modernist texts in the late nineteenth and early twentieth century are known for their experimentation with disjointed narratives, moving away from traditional linear narrative. Disjointed narratives are depicted in novels like *Catch 22*, where the author, Joseph Heller, structures the narrative based on free association of ideas rather than chronology. Another nonlinear narrative can be seen in the novel *Wuthering Heights*, written by Emily Bronte, which disrupts the chronological order by being told retrospectively after the first chapter. There seem to be two narratives in *Wuthering Heights* working at the same time: a present narrative as well as a past narrative. Authors employ disrupting narratives for various reasons; some use it for the purpose of creating situational irony for the readers, while some use it to create a certain effect in the reader, such as excitement, or even a feeling of discomfort or fear.

Sequence Structure in Technical Documents

The purpose of technical documents, such as instructions manuals, cookbooks, or "user-friendly" documents, is to provide information to users as clearly and efficiently as possible. In order to do this, the sequence structure in technical documents that should be used is one that is as straightforward as possible. This usually involves some kind of chronological order or a direct sequence of events. For example, someone who is reading an instruction manual on how to set up their Smart TV wants directions in a clear, simple, straightforward manner that does not leave the reader to guess at the proper sequence or lead to confusion.

Sequence Structure in Informational Texts

The structure in informational texts depends again on the genre. For example, a newspaper article may start by stating an exciting event that happened, and then move on to talk about that event in chronological order, known as sequence or order structure. Many informational texts also use cause and effect structure, which describes an event and then identifies reasons for why that event occurred. Some essays may write about their subjects by way of comparison and contrast, which is a structure that compares two things or contrasts them to highlight their differences. Other documents, such as proposals, will have a problem to solution structure, where the document highlights some kind of problem and then offers a solution toward the end. Finally, some informational texts are written with lush details and description in order to captivate the audience, allowing them to visualize the information presented to them. This type of structure is known as descriptive.

<u>*Making Inferences or Drawing Conclusions about Plots, Sequence of Events, Characters, Settings, and Ideas in Passages*</u>

One technique authors often use to make their fictional stories more interesting is not giving away too much information by providing hints and description. It is then up to the reader to draw a conclusion about the author's meaning by connecting textual clues with the reader's own pre-existing experiences and knowledge. Drawing conclusions is important as a reading strategy for understanding what is occurring in a text. Rather than directly stating who, what, where, when, or why, authors often describe story elements. Then, readers must draw conclusions to understand significant story components. As they go through a text, readers can think about the setting, characters, plot, problem, and solution; whether the author provided any clues for consideration; and combine any story clues with their existing knowledge and experiences to draw conclusions about what occurs in the text.

Making Predictions

Before and during reading, readers can apply the reading strategy of making predictions about what they think may happen next. For example, what plot and character developments will occur in fiction? What points will the author discuss in nonfiction? Making predictions about portions of text they have not yet read prepares readers mentally for reading, and also gives them a purpose for reading. To inform and make predictions about text, the reader can do the following:

- Consider the title of the text and what it implies
- Look at the cover of the book
- Look at any illustrations or diagrams for additional visual information
- Analyze the structure of the text
- Apply outside experience and knowledge to the text

Readers may adjust their predictions as they read. Reader predictions may or may not come true in text.

Making Inferences

Making an inference from a selection means to make an educated guess from the passage read. Inferences should be conclusions based off of sound evidence and reasoning. When multiple-choice test questions ask about the logical conclusion that can be drawn from reading text, the test-taker must identify which choice will unavoidably lead to that conclusion. In order to eliminate the incorrect choices, the test-taker should come up with a hypothetical situation wherein an answer choice is true, but the conclusion is not true.

For example, here is an example with three answer choices:

> Fred purchased the newest PC available on the market. Therefore, he purchased the most expensive PC in the computer store.
>
> What can one assume for this conclusion to follow logically?
>
> a. Fred enjoys purchasing expensive items.
> b. PCs are some of the most expensive personal technology products available.
> c. The newest PC is the most expensive one.

The premise of the text is the first sentence: Fred purchased the newest PC. The conclusion is the second sentence: Fred purchased the most expensive PC. Recent release and price are two different factors; the difference between them is the logical gap. To eliminate the gap, one must equate whatever new information the conclusion introduces with the pertinent information the premise has stated. This example simplifies the process by having only one of each: one must equate product recency with product price. Therefore, a possible bridge to the logical gap could be a sentence stating that the newest PCs always cost the most.

Analyzing Relationships within Passages, Including How People, Events, and Ideas are Connected
In a passage on the test, the relationships between people, events, and ideas may be clearly stated, or the reader might have to infer the relationships based on clues in the passage. To infer means to arrive at a conclusion based on evidence, clues, or facts.

People might be related through connections like family or friendship, or through events that link them directly or indirectly. In the passage, relationships may be described in background information or dialogue, or they may be implied through interactions between the characters.

Events and ideas in a passage can be related through sequence or through cause and effect. To relate events and ideas through sequence means to show what happened first and what happened next. Or a sequence could be ordered in another way, such as alphabetically, or geographically. Ideas and events can also be related through a comparison the author makes, showing the similarities and differences.

Sequence
When ideas are related through sequence, as in a series of events, the author typically uses signal words like *after, and then, while,* and *before.*

Cause and Effect
In cause and effect relationships, the cause is an event or circumstance that occurs before and is directly responsible for the effect. Signal words such as *because, due to,* and *as a result of* indicate a cause and effect relationship. When the relationship is implied, the reader must use clues in the passage to infer that the effect resulted from the cause.

Compare and Contrast
Ideas are often connected through comparisons of their similarities to one another using the signal words such as *like* or *and.* Ideas can also be connected through a contrast of their differences using signal words such as *but* or *however.*

Understanding Main Ideas and Details

Determining the Relationship Between Ideas
The main idea of a passage is related to the details that support it. These details demonstrate specific examples that support the main idea. For example, suppose a passage's main idea is "*Schools should be privatized*." The passage supports this main idea with this detail: "*Students who attend private schools perform better.*" The main idea is further supported by this statistical detail: "*65% of students with the highest national GPAs are from private schools.*"

Like the connections between events, these relationships may be clearly stated, or they may be implied. The reader can infer these relationships by analyzing signal words and the stated or implied connections. Below is a list of signal words.

Time Order
Time order can be inferred through words like *before* and *after,* or *next, then, first, lastly,* which signal that events occur one after the other.

Examples
Specific examples support a main idea or definitions.

Comparisons
The words like *and, also, like* signal a comparison of the similarities between ideas

Contrasts
Words like *but, however* signal a contrast of the differences between ideas.

Cause and Effect
Words like *because, so, therefore* signal that one event or circumstance is the cause of another.

Spatial
Ideas or objects are described by their location with respect to each other by using words such as *behind* and *to the right.*

Analyzing How Details Develop the Main Idea
A **paragraph** is made up of a series of related sentences that support a single main idea or message. Whether the main idea is clearly stated or implied by the supporting details, the reader can be guided to the desired conclusion.

The **main idea** is what the passage is about. When this main idea is clearly stated, the reader can easily determine how the evidence or details that follow support this main idea. Suppose the author states, "*Because of their inexperience with predators, dodos went extinct in the 17th century.*" The subsequent sentences will support this claim with details. The author will describe the sequence of events, each one leading the reader to agree with the statement based on the evidence and reach the conclusion the author wants them to reach.

If the main idea is not directly stated, the author must imply the main idea through the use of strong supporting details. These details could be comparisons of like ideas or contrasts of different ideas, factual evidence in graphs and statistics, quotes from experts, or vivid descriptions that evoke the desired emotional response.

Point of View and Purpose

Identifying the Author's Point of View and Purpose

Authors may have many purposes for writing a specific text. Their purposes may be to try and convince readers to agree with their position on a subject, to impart information, or to entertain. Other writers are motivated to write from a desire to express their own feelings. Authors' purposes are their reasons for writing something. A single author may have one overriding purpose for writing or multiple reasons. An author may explicitly state their intention in the text, or the reader may need to infer that intention. Those who read reflectively benefit from identifying the purpose because it enables them to analyze information in the text. By knowing why the author wrote the text, readers can glean ideas for how to approach it.

The following is a list of questions readers can ask in order to discern an author's purpose for writing a text:

- From the title of the text, why do you think the author wrote it?
- Was the purpose of the text to give information to readers?
- Did the author want to describe an event, issue, or individual?
- Was it written to express emotions and thoughts?
- Did the author want to convince readers to consider a particular issue?
- Was the author primarily motivated to write the text to entertain?
- Why do you think the author wrote this text from a certain point of view?
- What is your response to the text as a reader?
- Did the author state their purpose for writing it?

Students should read to interpret information rather than simply content themselves with roles as text consumers. Being able to identify an author's purpose efficiently improves reading comprehension, develops critical thinking, and makes students more likely to consider issues in depth before accepting writer viewpoints. Authors of fiction frequently write to entertain readers. Another purpose for writing fiction is making a political statement; for example, Jonathan Swift wrote "A Modest Proposal" (1729) as a political satire. Another purpose for writing fiction as well as nonfiction is to persuade readers to take some action or further a particular cause. Fiction authors and poets both frequently write to evoke certain moods; for example, Edgar Allan Poe wrote novels, short stories, and poems that evoke moods of gloom, guilt, terror, and dread. Another purpose of poets is evoking certain emotions: love is popular, as in Shakespeare's sonnets and numerous others. In "The Waste Land" (1922), T.S. Eliot evokes society's alienation, disaffection, sterility, and fragmentation.

Authors seldom directly state their purposes in texts. Students may be confronted with nonfiction texts such as biographies, histories, magazine and newspaper articles, and instruction manuals, among others. To identify the purpose in nonfiction texts, students can ask the following questions:

- Is the author trying to teach something?
- Is the author trying to persuade the reader?
- Is the author imparting factual information only?
- Is this a reliable source?
- Does the author have some kind of hidden agenda?

To help determine an author's purpose in nonfictional passages, students can also analyze sentence structure, word choice, and transitions to answer the aforementioned questions and to make inferences. For example, authors wanting to convince readers to view a topic negatively often choose words with negative connotations.

Narrative Writing

Narrative writing tells a story. The most prominent examples of narrative writing are fictional novels. Here are some examples:

- Mark Twain's The Adventures of Tom Sawyer and The Adventures of Huckleberry Finn
- Victor Hugo's *Les Misérables*
- Charles Dickens' Great Expectations, David Copperfield, and A Tale of Two Cities
- Jane Austen's Mansfield Park, Pride and Prejudice, Sense and Sensibility, and Emma
- Toni Morrison's Beloved, *The Bluest Eye*, and *Song of Solomon*
- Gabriel García Márquez's One Hundred Years of Solitude and Love in the Time of Cholera

Some nonfiction works are also written in narrative form. For example, some authors choose a narrative style to convey factual information about a topic, such as a specific animal, country, geographic region, and scientific or natural phenomenon.

Since narrative is the type of writing that tells a story, it must be told by someone, who is the narrator. The narrator may be a fictional character telling the story from their own viewpoint. This narrator uses the **first person** (*I, me, my, mine* and *we, us, our,* and *ours*). The narrator may simply be the author; for example, when Louisa May Alcott writes "Dear reader" in *Little Women*, she (the author) addresses us as readers. In this case, the novel is typically told in **third person**, referring to the characters as he, she, they, or them. Another more common technique is the **omniscient narrator**; i.e. the story is told by an unidentified individual who sees and knows everything about the events and characters—not only their externalized actions, but also their internalized feelings and thoughts. **Second person**, i.e. writing the story by addressing readers as "you" throughout, is less frequently used.

Expository Writing

Expository writing is also known as informational writing. Its purpose is not to tell a story as in narrative writing, to paint a picture as in descriptive writing, or to persuade readers to agree with something as in argumentative writing. Rather, its point is to communicate information to the reader. As such, the point of view of the author will necessarily be more objective. Whereas other types of writing appeal to the reader's emotions, appeal to the reader's reason by using logic, or use subjective descriptions to sway the reader's opinion or thinking, expository writing seeks to do none of these but simply to provide facts, evidence, observations, and objective descriptions of the subject matter. Some examples of expository writing include research reports, journal articles, articles and books about historical events or periods, academic subject textbooks, news articles and other factual journalistic reports, essays, how-to articles, and user instruction manuals.

Technical Writing

Technical writing is similar to expository writing in that it is factual, objective, and intended to provide information to the reader. Indeed, it may even be considered a subcategory of expository writing. However, technical writing differs from expository writing in that (1) it is specific to a particular field, discipline, or subject; and (2) it uses the specific technical terminology that belongs only to that area. Writing that uses technical terms is intended only for an audience familiar with those terms. A primary example of technical writing today is writing related to computer programming and use.

Persuasive Writing

Persuasive writing is intended to persuade the reader to agree with the author's position. It is also known as argumentative writing. Some writers may be responding to other writers' arguments, in which case they make reference to those authors or text and then disagree with them. However, another common technique is for the author to anticipate opposing viewpoints in general, both from other authors and from the author's own readers. The author brings up these opposing viewpoints, and then refutes them before they can even be raised, strengthening the author's argument. Writers persuade readers by appealing to their reason, which Aristotle called **logos;** appealing to emotion, which Aristotle called **pathos**; or appealing to readers based on the author's character and credibility, which Aristotle called **ethos.**

Determining How the Author Explains a Position and Responds to Different Viewpoints
An author should clearly state his or her opinion and use evidence, such as research studies, statistics, and examples, to support his or her stance. Although somewhat counterintuitive, raising opposing viewpoints or presenting the counterargument actually strengthens in argument (if done correctly). This is because it gives the author the platform to provide evidence against that viewpoint—to disprove it—instead of leaving readers wondering if another viewpoint is more rational.

Inferring the Author's Purpose in the Passage When it is Not Stated
The author's attitude toward a certain person or idea, or his or her purpose, may not always be stated. While it may seem impossible to know exactly what the author felt toward their subject, there are clues to indicate the emotion, or lack thereof, of the author. Clues like word choice or style will alert readers to the author's attitude. Some possible words that name the author's attitude are listed below:

- Admiring
- Angry
- Critical
- Defensive
- Enthusiastic
- Humorous
- Moralizing
- Neutral
- Objective
- Patriotic
- Persuasive
- Playful
- Sentimental
- Serious
- Supportive
- Sympathetic
- Unsupportive

An author's **tone** is the author's attitude toward their subject and is usually indicated by word choice. If an author's attitude toward their subject is one of disdain, the author will show the subject in a negative light, using deflating words or words that are negatively charged. If an author's attitude toward their subject is one of praise, the author will use agreeable words and show the subject in a positive light. If an author takes a neutral tone towards their subject, their words will be neutral as well, and they probably will show all sides of their subject, not just the negative or positive side.

Style is another indication of the author's attitude and includes aspects such as sentence structure, type of language, and formatting. Sentence structure is how a sentence is put together. Sometimes, short, choppy sentences will indicate a certain tone given the surrounding context, while longer sentences may serve to create a buffer to avoid being too harsh, or may be used to explain additional information. Style may also include formal or informal language. Using formal language to talk about a subject may indicate a level of respect. Using informal language may be used to create an atmosphere of friendliness or familiarity with a subject. Again, it depends on the surrounding context whether or not language is used in a negative or positive way. Style may also include formatting, such as determining the length of paragraphs or figuring out how to address the reader at the very beginning of the text.

The following is a passage from *The Florentine Painters of the Renaissance* by Bernhard Berenson. Following the passage is a question stem regarding the author's attitude toward their subject:

> Let us look now at an even greater triumph of movement than the Nudes, Pollaiuolo's "Hercules Strangling Antæus." As you realise the suction of Hercules' grip on the earth, the swelling of his calves with the pressure that falls on them, the violent throwing back of his chest, the stifling force of his embrace; as you realise the supreme effort of Antæus, with one hand crushing down upon the head and the other tearing at the arm of Hercules, you feel as if a fountain of energy had sprung up under your feet and were playing through your veins. I cannot refrain from mentioning still another masterpiece, this time not only of movement, but of tactile values and personal beauty as well—Pollaiuolo's "David" at Berlin. The young warrior has sped his stone, cut off the giant's head, and now he strides over it, his graceful, slender figure still vibrating with the rapidity of his triumph, expectant, as if fearing the ease of it. What lightness, what buoyancy we feel as we realise the movement of this wonderful youth!

Which one of the following best captures the author's attitude toward the paintings depicted in the passage?
   a. Neutrality towards the subject in this passage.
   b. Disdain for the violence found in the paintings.
   c. Excitement for the physical beauty found within the paintings.
   d. Passion for the movement and energy of the paintings.
   e. Seriousness for the level of artistry the paintings hold.

Choice *D* is the best answer. We know that the author feels positively about the subject because of the word choice. Berenson uses words and phrases like "supreme," "fountain of energy," "graceful," "figure still vibrating," "lightness," "buoyancy," and "wonderful youth." Notice, also, the exclamation mark at the end of the paragraph. These words and style depict an author full of passion, especially for the movement and energy found within the paintings.

Choice *A* is incorrect because the author is biased towards the subject due to the energy he writes with— he calls the movement in the paintings "wonderful" and by the other word choices and phrases, readers can tell that this is not an objective analysis of these paintings. Choice *B* is incorrect because, although the author does mention the "violence" in the stance of Hercules, he does not exude disdain towards this. Choice *C* is incorrect. There is excitement in the author's tone, and some of this excitement is directed towards the paintings' physical beauty. However, this is not the *best* answer choice. Choice *D* is more accurate when stating the passion is for the movement and energy of the paintings, of which physical beauty is included. Finally, Choice *E* is incorrect. The tone is partly serious, but we see the author getting carried away with enthusiasm for the beauty of the paintings towards the middle and especially the end of the passage.

Tone and Figurative Language

Understanding How Words Affect Tone
Words can be very powerful. When written words are used with the intent to make an argument or support a position, the words used—and the way in which they are arranged—can have a dramatic effect

on the readers. Clichés, colloquialisms, run-on sentences, and misused words are all examples of ways that word choice can negatively affect writing quality. Unless the writer carefully considers word choice, a written work stands to lose credibility.

If a writer's overall intent is to provide a clear meaning on a subject, he or she must consider not only the exact words to use, but also their placement, repetition, and suitability. Academic writing should be intentional and clear, and it should be devoid of awkward or vague descriptions that can easily lead to misunderstandings. When readers find themselves reading and rereading just to gain a clear understanding of the writer's intent, there may be an issue with word choice. Although the words used in academic writing are different from those used in a casual conversation, they shouldn't necessarily be overly academic either. It may be relevant to employ key words that are associated with the subject, but struggling to inject these words into a paper just to sound academic may defeat the purpose. If the message cannot be clearly understood the first time, word choice may be the culprit.

Word choice also conveys the author's attitude and sets a tone. Although each word in a sentence carries a specific **denotation**, it might also carry positive or negative **connotations**—and it is the connotations that set the tone and convey the author's attitude. Consider the following similar sentences:

> It was the same old routine that happens every Saturday morning—eat, exercise, chores.

> The Saturday morning routine went off without a hitch—eat, exercise, chores.

The first sentence carries a negative connotation with the author's "same old routine" word choice. The feelings and attitudes associated with this phrase suggest that the author is bored or annoyed at the Saturday morning routine. Although the second sentence carries the same topic—explaining the Saturday morning routine—the choice to use the expression "without a hitch" conveys a positive or cheery attitude.

An author's writing style can likewise be greatly affected by word choice. When writing for an academic audience, for example, it is necessary for the author to consider how to convey the message by carefully considering word choice. If the author interchanges between third-person formal writing and second-person informal writing, the author's writing quality and credibility are at risk. Formal writing involves complex sentences, an objective viewpoint, and the use of full words as opposed to the use of a subjective viewpoint, contractions, and first- or second-person usage commonly found in informal writing.

Content validity, the author's ability to support the argument, and the audience's ability to comprehend the written work are all affected by the author's word choice.

Understanding How Figurative Language Affects the Meaning of Words or Phrases
Authors of a text use language with multiple levels of meaning for many different reasons. When the meaning of a text calls for directness, literal language should be used to provide clarity to the reader. Figurative language can be used when the author wants to produce an emotional effect in the reader or facilitate a deeper understanding of a word or passage. For example, if someone wanted to write a set of instructions on how to use a computer, they would write in literal language. However, if someone wanted to comment on the social implications of banning immigration, they might want to use a wide range of figurative language to highlight an empathetic response. It is important to keep in mind, too, that a single text can have a mixture of both literal and figurative language.

Literal Language
Literal language uses words in accordance with their actual definition. Many informational texts employ literal language because it is straightforward and precise. Documents such as instructions, proposals,

technical documents, and workplace documents use literal language for the majority of their writing, so there is no confusion or complexity of meaning for readers to decipher. The information is best communicated through clear and precise language.

The following are brief examples of literal language:

- I cook with olive oil.
- There are 365 days in a year.
- My grandma's name is Barbara.
- Yesterday we had some scattered thunderstorms.
- World War II began in 1939.
- Blue whales are the largest species of whale.

Figurative Language

Not meant to be taken literally, figurative language is useful when the author of a text wants to produce an emotional effect in the reader or add a heightened complexity to the meaning of the text. Figurative language is used more heavily in texts such as literary fiction, poetry, critical theory, and speeches. Figurative language goes beyond literal language, allowing readers to form associations they wouldn't normally form with literal language. Using language in a figurative sense appeals to the imagination of the reader. It is important to remember that words themselves are signifiers of objects and ideas, and not the objects and ideas themselves. Figurative language can highlight this detachment by creating multiple associations, but also points to the fact that language is fluid and capable of creating a world full of linguistic possibilities. Figurative language, it can be argued, is the heart of communication even outside of fiction and poetry. People connect through humor, metaphors, cultural allusions, puns, and symbolism in their everyday rhetoric. The following are terms associated with figurative language:

A **simile** is a comparison of two things using *like, than,* or *as*. A simile usually takes objects that have no apparent connection, such as a mind and an orchid, and compares them:

> His mind was as complex and rare as a field of ghost orchids.

Similes encourage a new, fresh perspective on objects or ideas that wouldn't otherwise occur. Similes are different than metaphors, which also make comparisons. However, **metaphors** do not use *like, than,* or *as*. So, a metaphor from the above example would be:

> His mind was a field of ghost orchids.

Thus, similes highlight the comparison by focusing on the figurative side of the language, elucidating more the author's intent: a field of ghost orchids is something complex and rare, like the mind of a genius. With the metaphor, however, we get a beautiful yet somewhat equivocal comparison.

A popular use of figurative language, metaphors compare objects or ideas directly, asserting that something *is* a certain thing, even if it isn't. The following is an example of a metaphor used by writer Virginia Woolf:

> Books are the mirrors of the soul.

Metaphors have a vehicle and a tenor. The tenor is "books" and the vehicle is "mirrors of the soul." That is, the **tenor** is what is meant to be described, and the **vehicle** is that which carries the weight of the comparison. In this metaphor, perhaps the author means to say that written language (books) reflect a person's most inner thoughts and desires.

There are also **dead metaphors**, which means that the phrases have been so overused to the point where the figurative meaning becomes literal, like the phrase "What you're saying is crystal clear." The phrase compares "what's being said" to something "crystal clear." However, since the latter part of the phrase is in such popular use, the meaning seems literal ("I understand what you're saying") even when it's not.

Finally, an **extended metaphor** is a metaphor that goes on for several paragraphs, or even an entire text. John Keats' poem "On First Looking into Chapman's Homer" begins, "Much have I travell'd in the realms of gold," and goes on to explain the first time he hears Chapman's translation of Homer's writing. We see the extended metaphor begin in the first line. Keats is comparing travelling into "realms of gold" and exploration of new lands to the act of hearing a certain kind of literature for the first time. The extended metaphor goes on until the end of the poem where Keats stands "Silent, upon a peak in Darien," having heard the end of Chapman's translation. Keats has gained insight into new lands (new text) and is the richer for it.

The following are brief definitions and examples of other common types figurative language:

- **Onomatopoeia**: A word that, when spoken, imitates the sound to which it refers. Ex: "We heard a loud *boom* while driving to the beach yesterday."

- **Personification**: When human characteristics are given to animals, inanimate objects, or abstractions. An example would be in William Wordsworth's poem "Daffodils" where he sees a "crowd . . . / of golden daffodils . . . / Fluttering and dancing in the breeze." Dancing is usually a characteristic attributed solely to humans, but Wordsworth personifies the daffodils here as a crowd of people dancing.

- **Juxtaposition**: Juxtaposition is placing two objects side by side for comparison. In literature, this might look like placing two characters side by side for contrasting effect, like God and Satan in Milton's "Paradise Lost."

- **Paradox**: A paradox is a statement that is self-contradictory but will be found nonetheless true. One example of a paradoxical phrase is when Socrates said "I know one thing; that I know nothing." Seemingly, if Socrates knew nothing, he wouldn't know that he knew nothing. However, it is one thing he knows: that true wisdom begins with casting all presuppositions one has about the world aside.

- **Hyperbole**: A hyperbole is an exaggeration. Ex: "I'm so tired I could sleep for centuries."

- **Allusion**: An allusion is a reference to a character or event that happened in the past. An example of a poem littered with allusions is T.S. Eliot's "The Waste Land." An example of a biblical allusion manifests when the poet says, "I will show you fear in a handful of dust," creating an ominous tone from Genesis 3:19 "For you are dust, and to dust you shall return."

- **Pun**: Puns are used in popular culture to invoke humor by exploiting the meanings of words. They can also be used in literature to give hints of meaning in unexpected places. One example of a pun is when Mercutio is giving his monologue after he is stabbed by Tybalt in "Romeo and Juliet" and says, "look for me tomorrow and you will find me a grave man."

- **Imagery**: This is a collection of images given to the reader by the author. If a text is rich in imagery, it is easier for the reader to imagine themselves in the author's world.

  One example of a poem that relies on imagery is William Carlos Williams' "The Red Wheelbarrow":

  > so much depends
  > upon
  >
  > a red wheel
  > barrow
  >
  > glazed with rain
  > water
  >
  > beside the white
  > chickens

  The starkness of the imagery and the placement of the words in the poem, to some readers, throw the poem into a meditative state where, indeed, the world of this poem is made up solely of images of a purely simple life. This poem tells a story in sixteen words by using imagery.

- **Symbolism**: A **symbol** is used to represent an idea or belief system. For example, poets in Western civilization have been using the symbol of a rose for hundreds of years to represent love. In Japan, poets have used the firefly to symbolize passionate love, and sometimes even spirits of those who have died. Symbols can also express powerful political commentary and can be used in propaganda.

- **Irony**: There are three types of irony. **Verbal** irony is when a person states one thing and means the opposite. For example, a person is probably using irony when they say, "I can't wait to study for this exam next week." **Dramatic** irony occurs in a narrative and happens when the audience knows something that the characters do not. In the modern TV series *Hannibal*, we as an audience know that Hannibal Lecter is a serial killer, but most of the main characters do not. This is dramatic irony. Finally, **situational** irony is when one expects something to happen, and the opposite occurs. For example, we can say that a fire station burning down would be an instance of situational irony.

Understanding How the Use of Words, Phrases, or Figurative Language Influences the Author's Purpose
As mentioned, **denotation** refers to a word's explicit definition, like that found in the dictionary. Denotation is often set in comparison to connotation. **Connotation** is the emotional, cultural, social, or personal implication associated with a word. Denotation is more of an objective definition, whereas connotation can be more subjective, although many connotative meanings of words are similar for certain cultures. The denotative meanings of words are usually based on facts, and the connotative meanings of words are usually based on emotion.

Here are some examples of words and their denotative and connotative meanings in Western culture:

Word	Denotative Meaning	Connotative Meaning
Home	A permanent place where one lives, usually as a member of a family.	A place of warmth; a place of familiarity; comforting; a place of safety and security. "Home" usually has a positive connotation.
Snake	A long reptile with no limbs and strong jaws that moves along the ground; some snakes have a poisonous bite.	An evil omen; a slithery creature (human or nonhuman) that is deceitful or unwelcome. "Snake" usually has a negative connotation.
Winter	A season of the year that is the coldest, usually from December to February in the northern hemisphere and from June to August in the southern hemisphere.	Circle of life, especially that of death and dying; cold or icy; dark and gloomy; hibernation, sleep, or rest. Winter can have a negative connotation, although many who have access to heat may enjoy the snowy season from their homes.

Organizing Ideas

Determining How a Section Fits into a Passage and Helps Develop the Ideas
To determine whether details support the main idea of a passage, determine the relationships between ideas. For example, suppose the main idea of the passage is _"Due to inflation, the value of a dollar is different today than it was 100 years ago and different than it will be 100 years from now."_ This idea is then directly supported by the statements like this: _"In 1918, an item that cost a dollar would cost almost $17 today."_ Another detail supporting it is this: _"Experts estimate that the expected rate of inflation over the next 10 years is about 2%."_

To determine whether a detail supports the main idea, a reader must determine whether there is a connection. Then they must decide whether the detail directly supports the main idea, one of the supporting ideas, or none of the ideas. Suppose a supporting idea in a passage about inflation is _"In 1918, an item that cost a dollar would cost almost $17 today."_ The reader then reads this detail: _"That means a coffee that costs $5 today would have cost about 30 cents back then."_ The detail about the coffee price is an example of the supporting idea about prices in 1918. However, the statement _The value of an item is determined by how popular the item is_ does not support the main idea (_"Due to inflation, the value of a dollar is different today than it was 100 years ago and different than it will be 100 years from now."_) or any of the supporting ideas.

Analyzing How a Text is Organized

Text structure is the way in which the author organizes and presents textual information so readers can follow and comprehend it. One kind of text structure is sequence. This means the author arranges the text in a logical order from beginning to middle to end. There are three types of sequences:

- **Chronological**: ordering events in time from earliest to latest

- **Spatial**: describing objects, people, or spaces according to their physical relationships to one another in space

- **Order of Importance**: addressing topics, characters, or ideas according to how important they are, from either least important to most important

Chronological sequence is the most common sequential text structure. Readers can identify sequential structure by looking for words that signal it, like *first, earlier, meanwhile, next, then, later, finally;* and specific times and dates the author includes as chronological references.

Problem-Solution Text Structure

The problem-solution text structure organizes textual information by presenting readers with a problem and then developing its solution throughout the course of the text. The author may present a variety of alternatives as possible solutions, eliminating each as they are found unsuccessful, or gradually leading up to the ultimate solution. For example, in fiction, an author might write a murder mystery novel and have the character(s) solve it through investigating various clues or character alibis until the killer is identified. In nonfiction, an author writing an essay or book on a real-world problem might discuss various alternatives and explain their disadvantages or why they would not work before identifying the best solution. For scientific research, an author reporting and discussing scientific experiment results would explain why various alternatives failed or succeeded.

Comparison-Contrast Text Structure

Comparison identifies similarities between two or more things. **Contrast** identifies differences between two or more things. Authors typically employ both to illustrate relationships between things by highlighting their commonalities and deviations. For example, a writer might compare Windows and Linux as operating systems, and contrast Linux as free and open-source vs. Windows as proprietary. When writing an essay, sometimes it is useful to create an image of the two objects or events you are comparing or contrasting. Venn diagrams are useful because they show the differences as well as the similarities between two things. Once you've seen the similarities and differences on paper, it might be helpful to create an outline of the essay with both comparison and contrast. Every outline will look different, because every two or more things will have a different number of comparisons and contrasts. Say you are trying to compare and contrast carrots with sweet potatoes.

Here is an example of a compare/contrast outline using those topics:

- **Introduction:** Talk about why you are comparing and contrasting carrots and sweet potatoes. Give the thesis statement.
- **Body paragraph 1:** Sweet potatoes and carrots are both root vegetables (similarity)
- **Body paragraph 2:** Sweet potatoes and carrots are both orange (similarity)
- **Body paragraph 3:** Sweet potatoes and carrots have different nutritional components (difference)
- **Conclusion:** Restate the purpose of your comparison/contrast essay.

Of course, if there is only one similarity between your topics and two differences, you will want to rearrange your outline. Always tailor your essay to what works best with your topic.

Descriptive Text Structure
Description can be both a type of text structure and a type of text. Some texts are descriptive throughout entire books. For example, a book may describe the geography of a certain country, state, or region, or tell readers all about dolphins by describing many of their characteristics. Many other texts are not descriptive throughout, but use descriptive passages within the overall text. The following are a few examples of descriptive text:

- When the author describes a character in a novel
- When the author sets the scene for an event by describing the setting
- When a biographer describes the personality and behaviors of a real-life individual
- When a historian describes the details of a particular battle within a book about a specific war
- When a travel writer describes the climate, people, foods, and/or customs of a certain place

A hallmark of description is using sensory details, painting a vivid picture so readers can imagine it almost as if they were experiencing it personally.

Cause and Effect Text Structure
When using cause and effect to extrapolate meaning from text, readers must determine the cause when the author only communicates effects. For example, if a description of a child eating an ice cream cone includes details like beads of sweat forming on the child's face and the ice cream dripping down her hand faster than she can lick it off, the reader can infer or conclude it must be hot outside. A useful technique for making such decisions is wording them in "If...then" form, e.g. "If the child is perspiring and the ice cream melting, it may be a hot day." Cause and effect text structures explain why certain events or actions resulted in particular outcomes. For example, an author might describe America's historical large flocks of dodo birds, the fact that gunshots did not startle/frighten dodos, and that because dodos did not flee, settlers killed whole flocks in one hunting session, explaining how the dodo was hunted into extinction.

Understanding the Meaning and Purpose of Transition Words
In connected writing, some sentences naturally lead to others, whereas in other cases, a new sentence expresses a new idea. We use transitional phrases to connect sentences and the ideas they convey. This makes the writing coherent. Transitional language also guides the reader from one thought to the next. For example, when pointing out an objection to the previous idea, starting a sentence with "However," "But," or "On the other hand" is transitional. When adding another idea or detail, writers use "Also," "In addition," "Furthermore," "Further," "Moreover," "Not only," etc. Readers have difficulty perceiving connections between ideas without such transitional wording.

Analyzing How the Organization of a Paragraph or Passage Supports the Author's Ideas
The author's purpose determines the organization of a passage. The organization guides the connections between the ideas that support the main idea and establishes relationships between supporting ideas.

For example, suppose an author wants to explain why Henry VIII created the Church of England. This author might choose a sequential structure with time order relationships to describe the events. She would also include the cause and effect structure between the ideas to establish the relationship between the result and the circumstances that caused the result: *"Henry VIII created the Church of England because he was excommunicated from the Catholic church."*

If the author's purpose is to convince someone to leave the Church of England to return to Catholicism, they might use comparison and contrast to describe the ideals of the two churches. The same series of events and cause and effect relationships contribute to the main idea, but the conclusion the author wants the reader to make is different: *"The foundation of the Church of England is sinful."* The author could also use order of importance to provide a list of reasons why this event occurred, or the reasons the reader should leave the Church of England.

Comparing Different Ways of Presenting Ideas

Evaluating Two Different Texts and How They Address Scope, Purpose, Emphasis, Audience, and Impact
The author's purpose for writing the passage guides how they address the audience, the scope or level of detail of the passage, what to emphasize, and the impact they want the passage to have on the reader.

If the author wants to persuade an audience of voters to vote for an independent candidate, they might show parallels between the sentiments that divided the North and South leading up to the Civil War and the division between the two major parties in the US government today. The author would emphasize the opposing views of the two sides in the Civil War and their inability to compromise, and the author would demonstrate a similar lack of compromise in politics today. The author might compare the issues that divided the sides and the possible reasons behind each side's position, such as the economic power of the voters or human rights, and they might emphasize the outcome of the conflict then. The success of or failure of the author to convince their audience to vote for the independent candidate would determine the impact of the passage.

Using the same information, a passage about voting and the Civil War could be used to persuade voters to vote for one of the two major parties. A passage like that would have a different scope, emphasis, and impact than one about independent candidates. The author in favor of a major party would focus the scope of the argument on different details, drawing positive comparisons to their party and negative comparisons to the opposing party. The author might emphasize the result of the conflict, rather than the events that led to war; or the author might minimize the impact of the division between the two sides. This author might emphasize the great deeds the party has accomplished, despite the opposition. The success of this author in convincing voters to back a particular side would determine the impact of the passage.

Evaluating Point of View, Tone, Style, Organization, Purpose, or Impact in Two Different Passages
The author's purpose affects the way they address the point of view from which they are writing, the tone and style of the passage, and the organization they use.

Suppose a passage has this main idea: *"Nuclear power is safe."* If the author is an investor in nuclear power plant construction, their purpose might be change the public's perception of nuclear power and smooth the way for any permits. The author might use a persuasive style and a friendly tone to convince people to feel good about nuclear power.

Suppose a passage has this main idea: *"All high school graduates should go to college because it's the way to get ahead in life."* If the author is the dean of students at a college, he might believe in college for and also believe his college is the best. His secondary purpose might be to recruit more students to attend his college. From his point of view, students should attend his school. He might use a persuasive style and a friendly tone to convince graduates using examples and testimonials from students that his college offers the best academics and the most exciting campus life.

Lastly, suppose a passage were written from the point of view of a low-income high school graduate who can't afford college. This passage's main idea might be: *"All students should be able to attend college."* Instead of potential college students, the audience for this passage might be a state Congressperson, and the author's purpose might be to advocate better funding for state universities. The style of this passage might also include persuasive language. The tone would be serious, and the author might use statistics to contrast the lives of those who attended college and those who did not.

Humanities

The Relationship Between Evidence and Main Ideas and Details

Summarizing Information from a Passage

An important skill is the ability to read a complex text and then reduce its length and complexity by focusing on the key events and details. A **summary** is a shortened version of the original text, written by the reader in their own words. The summary should be shorter than the original text, and it must be thoughtfully formed to include critical points from the original text.

In order to effectively summarize a complex text, it's necessary to understand the original source and identify the major points covered. It may be helpful to outline the original text to get the big picture and avoid getting bogged down in the minor details. For example, a summary wouldn't include a statistic from the original source unless it was the major focus of the text. It's also important for readers to use their own words, yet retain the original meaning of the passage. The key to a good summary is emphasizing the main idea without changing the focus of the original information.

The more complex a text, the more difficult it can be to summarize. Readers must evaluate all points from the original source and then filter out what they feel are the less necessary details. Only the essential ideas should remain. The summary often mirrors the original text's organizational structure. For example, in a problem-solution text structure, the author typically presents readers with a problem and then develops solutions through the course of the text. An effective summary would likely retain this general structure, rephrasing the problem and then reporting the most useful or plausible solutions.

Paraphrasing is somewhat similar to summarizing. It calls for the reader to take a small part of the passage and list or describe its main points. Paraphrasing is more than rewording the original passage, though. Like summary, it should be written in the reader's own words, while still retaining the meaning of the original source. The main difference between summarizing and paraphrasing is that a summary would be appropriate for a much larger text, while paraphrase might focus on just a few lines of text. Effective paraphrasing will indicate an understanding of the original source, yet still help the reader expand on their interpretation. A paraphrase should neither add new information nor remove essential facts that change the meaning of the source.

Identifying the Relationship Between the Main Idea and Details of a Passage

The **topic** of a text is the general subject matter. Text topics can usually be expressed in one word, or a few words at most. Additionally, readers should ask themselves what point the author is trying to make. This point is the **main idea** of the text, the one thing the author wants readers to know concerning the topic. Once the author has established the main idea, they will support the main idea by supporting details. **Supporting details** are evidence that support the main idea and include personal testimonies, examples, or statistics.

One analogy for these components and their relationships is that a text is like a well-designed house. The topic is the roof, covering all rooms. The main idea is the frame. The supporting details are the various rooms. To identify the topic of a text, readers can ask themselves what or who the author is writing about in the paragraph. To locate the main idea, readers can ask themselves what one idea the author wants readers to know about the topic. To identify supporting details, readers can put the main idea into question form and ask "what does the author use to prove or explain their main idea?"

Let's look at an example. An author is writing an essay about the Amazon rainforest and trying to convince the audience that more funding should go into protecting the area from deforestation. The author makes the argument stronger by including evidence of the benefits of the rainforest: it provides habitats to a variety of species, it provides much of the earth's oxygen which in turn cleans the atmosphere, and it is the home to medicinal plants that may be the answer to some of the world's deadliest diseases. Here is an outline of the essay looking at topic, main idea, and supporting details:

- Topic: Amazon rainforest
- Main Idea: The Amazon rainforest should receive more funding to protect it from deforestation.
- Supporting Details:
  1. It provides habitats to a variety of species
  2. It provides much of the earth's oxygen which in turn cleans the atmosphere
  3. It is home to medicinal plants that may be the answer to some of the deadliest diseases.

Notice that the topic of the essay is listed in a few key words: "Amazon rainforest." The main idea tells us what about the topic is important: that the topic should be funded in order to prevent deforestation. Finally, the supporting details are what author relies on to convince the audience to act or to believe in the truth of the main idea.

Determining the Main idea of a Passage
The main idea a paragraph is sometimes clearly stated at the beginning or the end of the paragraph, or it might be implied so the reader must infer it from supporting details.

To determine the main idea of a paragraph, the reader must determine the topic. The topic is different than the main idea. The main idea is the author's message; the topic is who or what the passage is about. To determine the topic when it is not explicitly stated, the reader examines the supporting details within the paragraph.

These details might include answers to questions raised throughout the passage, events or ideas that move a story along, or evidence that leads the reader to a specific conclusion. These details might include vivid descriptions of events, quotes that offer an opinion, comparisons of ideas, and statistics and graphs.

Based on the topic and the supporting details or evidence, the reader can form a conclusion about the author's purpose to infer the main idea of the paragraph.

Determining Which Details Support a Main Idea
Authors use both major and minor details to support the main idea of a passage.

Major details are general statements that contribute directly to supporting the main idea. For example, suppose the main idea of a passage is "*Many variables determine what type of car you should buy.*" The following major detail supports this main idea: "*One of the variables is where you live.*"

Minor details are specific examples of the major details. For example, consider these details:

If it snows a lot, you might need AWD. If you live in the city and have to commute, you might want good gas mileage.

These are minor details that support one of the passage's major details: *One of the variables is where you live.* Minor details contribute to the overall message of the main idea, but they do not directly support the main idea. Instead, they develop and support the major details that support the main idea.

Identifying the Theme and Supportive Elements in Fiction and Nonfiction
The **theme** of a piece of text is the central idea the author communicates. Whereas the topic of a passage of text may be concrete in nature, by contrast the theme is always conceptual. For example, while the topic of Mark Twain's novel *The Adventures of Huckleberry Finn* might be described as something like the coming-of-age experiences of a poor, illiterate, functionally orphaned boy around and on the Mississippi River in 19th-century Missouri, one theme of the book might be that human beings are corrupted by society. Another might be that slavery and "civilized" society itself are hypocritical. Whereas the main idea in a text is the most important single point that the author wants to make, the theme is the concept or view around which the author centers the text.

Throughout time, humans have told stories with similar themes. Some themes are universal across time, space, and culture. These include themes of the individual as a hero, conflicts of the individual against nature, the individual against society, change vs. tradition, the circle of life, coming-of-age, and the complexities of love. Themes involving war and peace have featured prominently in diverse works, like Homer's *Iliad*, Tolstoy's *War and Peace* (1869), Stephen Crane's *The Red Badge of Courage* (1895), Hemingway's *A Farewell to Arms* (1929), and Margaret Mitchell's *Gone with the Wind* (1936). Another universal literary theme is that of the quest. These appear in folklore from countries and cultures worldwide, including the Gilgamesh Epic, Arthurian legend's Holy Grail quest, Virgil's *Aeneid*, Homer's *Odyssey*, and the *Argonautica*. Cervantes' *Don Quixote* is a parody of chivalric quests. J.R.R. Tolkien's *The Lord of the Rings* trilogy (1954) also features a quest.

One instance of similar themes across cultures is when those cultures are in countries that are geographically close to each other. For example, a folklore story of a rabbit in the moon using a mortar and pestle is shared among China, Japan, Korea, and Thailand—making medicine in China, making rice cakes in Japan and Korea, and hulling rice in Thailand. Another instance is when cultures are more distant geographically, but their languages are related. For example, East Turkestan's Uighurs and people in Turkey share tales of folk hero Effendi Nasreddin Hodja. Another instance, which may either be called cultural diffusion or simply reflect commonalities in the human imagination, involves shared themes among geographically and linguistically different cultures: both Cameroon's and Greece's folklore tell of centaurs; Cameroon, India, Malaysia, Thailand, and Japan, of mermaids; Brazil, Peru, China, Japan, Malaysia, Indonesia, and Cameroon, of underwater civilizations; and China, Japan, Thailand, Vietnam, Malaysia, Brazil, and Peru, of shape-shifters.

Two prevalent literary themes are love and friendship, which can end happily, sadly, or both. William Shakespeare's *Romeo and Juliet*, Emily Brontë's *Wuthering Heights*, Leo Tolstoy's *Anna Karenina*, and both *Pride and Prejudice* and *Sense and Sensibility* by Jane Austen are famous examples. Another theme recurring in popular literature is of revenge, an old theme in dramatic literature, e.g. Elizabethans Thomas Kyd's *The Spanish Tragedy* and Thomas Middleton's *The Revenger's Tragedy*. Some more well-known instances include Shakespeare's tragedies *Hamlet* and *Macbeth*, Alexandre Dumas' *The Count of Monte Cristo*, John Grisham's *A Time to Kill*, and Stieg Larsson's *The Girl Who Kicked the Hornet's Nest*.

Themes are underlying meanings in literature. For example, if a story's main idea is a character succeeding against all odds, the theme is overcoming obstacles. If a story's main idea is one character wanting what another character has, the theme is jealousy. If a story's main idea is a character doing something they were afraid to do, the theme is courage. Themes differ from topics in that a topic is a subject matter; a theme is the author's opinion about it. For example, a work could have a topic of war and a theme that war is a curse. Authors present themes through characters' feelings, thoughts, experiences, dialogue, plot actions, and events. Themes function as "glue" holding other essential story elements together. They offer readers insights into characters' experiences, the author's philosophy, and how the world works.

Drawing Conclusions, Making Inferences, and Evaluating Evidence

Making Generalizations Based on Evidence
Readers form hypotheses about the main idea as they read a passage. Hypotheses are based on incomplete evidence, so as the reader gathers more the details from the passage, their hypothesis changes or is proven by the evidence. They start to form generalizations about the author's message.

Generalizations are broad conclusions based on the way the reader interprets the details. These generalizations might be correct, or the reader might have made inferences from the evidence that the author did not intend.

To determine if these generalizations were implied by the author, or if they are incorrect, the reader should examine the evidence they collected.

For example:

- Humans cannot live without water.
- People use thousands of gallons of water a year to keep their lawns green.

Based on these details, the reader concludes that *people who water their lawns are wasting water.*

The author has not offered evidence that watering lawns is wasteful, so the generalization is not supported. If the passage includes the detail *scientists say water will soon be dangerously scarce*, the generalization is supported. However, the following statement would contradict the reader's generalization: *"Lawns also need water to grow and sustain life."*

Using Main Ideas to Draw Conclusions
Authors describe settings, characters, character emotions, and events. Readers must infer to understand text fully. Inferring enables readers to figure out meanings of unfamiliar words, make predictions about upcoming text, draw conclusions, and reflect on reading. Readers can infer about text before, during, and after reading. In everyday life, we use sensory information to infer. Readers can do the same with text. When authors do not answer all reader questions, readers must infer by saying "I think….This could be….This is because….Maybe….This means….I guess…" etc. Looking at illustrations, considering characters' behaviors, and asking questions during reading facilitate inference. Taking clues from text and connecting text to prior knowledge help to draw conclusions. Readers can infer word meanings, settings, reasons for occurrences, character emotions, pronoun referents, author messages, and answers to questions unstated in text. To practice inference, students can read sentences written/selected by the instructor, discuss the setting and character, draw conclusions, and make predictions.

Making inferences and drawing conclusions involve skills that are quite similar: both require readers to fill in information the author has omitted. Authors may omit information as a technique for inducing readers to discover the outcomes themselves; or they may consider certain information unimportant; or they may assume their reading audience already knows certain information. To make an inference or draw a conclusion about text, readers should observe all facts and arguments the author has presented and consider what they already know from their own personal experiences. Reading students taking multiple-choice tests that refer to text passages can determine correct and incorrect choices based on the information in the passage. For example, from a text passage describing an individual's signs of anxiety while unloading groceries and nervously clutching their wallet at a grocery store checkout, readers can infer or conclude that the individual may not have enough money to pay for everything.

Describing the Steps of an Argument

When authors write text for the purpose of persuading others to agree with them, they assume a position with the subject matter about which they are writing. Rather than presenting information objectively, the author treats the subject matter subjectively so that the information presented supports his or her position. In their argumentation, the author presents information that refutes or weakens opposing positions. Another technique authors use in persuasive writing is to anticipate arguments against the position. When students learn to read subjectively, they gain experience with the concept of persuasion in writing, and learn to identify positions taken by authors. This enhances their reading comprehension and develops their skills for identifying pro and con arguments and biases.

There are five main parts of the classical argument that writers employ in a well-designed stance:

- **Introduction:** In the introduction to a classical argument, the author establishes goodwill and rapport with the reading audience, warms up the readers, and states the thesis or general theme of the argument.

- **Narration:** In the narration portion, the author gives a summary of pertinent background information, informs the readers of anything they need to know regarding the circumstances and environment surrounding and/or stimulating the argument, and establishes what is at risk or the stakes in the issue or topic. Literature reviews are common examples of narrations in academic writing.

- **Confirmation:** The confirmation states all claims supporting the thesis and furnishes evidence for each claim, arranging this material in logical order—e.g. from most obvious to most subtle or strongest to weakest.

- **Refutation and Concession:** The refutation and concession discuss opposing views and anticipate reader objections without weakening the thesis, yet permitting as many oppositions as possible.

- **Summation:** The summation strengthens the argument while summarizing it, supplying a strong conclusion and showing readers the superiority of the author's solution.

Introduction

A classical argument's introduction must pique reader interest, get readers to perceive the author as a writer, and establish the author's position. Shocking statistics, new ways of restating issues, or quotations or anecdotes focusing the text can pique reader interest. Personal statements, parallel instances, or analogies can also begin introductions—so can bold thesis statements if the author believes readers will agree. Word choice is also important for establishing author image with readers.

The introduction should typically narrow down to a clear, sound thesis statement. If readers cannot locate one sentence in the introduction explicitly stating the writer's position or the point they support, the writer probably has not refined the introduction sufficiently.

Narration and Confirmation

The narration part of a classical argument should create a context for the argument by explaining the issue to which the argument is responding, and by supplying any background information that influences the issue. Readers should understand the issues, alternatives, and stakes in the argument by the end of the narration to enable them to evaluate the author's claims equitably. The confirmation part of the classical argument enables the author to explain why they believe in the argument's thesis. The author builds a chain of reasoning by developing several individual supporting claims and explaining why that evidence supports each claim and also supports the overall thesis of the argument.

Refutation and Concession and Summation

The classical argument is the model for argumentative/persuasive writing, so authors often use it to establish, promote, and defend their positions. In the refutation aspect of the refutation and concession part of the argument, authors disarm reader opposition by anticipating and answering their possible objections, persuading them to accept the author's viewpoint. In the concession aspect, authors can concede those opposing viewpoints with which they agree.

This can avoid weakening the author's thesis while establishing reader respect and goodwill for the author: all refutation and no concession can antagonize readers who disagree with the author's position. In the conclusion part of the classical argument, a less skilled writer might simply summarize or restate the thesis and related claims; however, this does not provide the argument with either momentum or closure. More skilled authors revisit the issues and the narration part of the argument, reminding readers of what is at stake.

Identifying Evidence Used to Support a Claim or Conclusion

On the Reading section of the SSAT there will likely be questions that ask the test taker about evidence or principles expressed in the selection that pertain to the argument. A **principle** functions as a fundamental truth used as a basis for a scenario or system of reasoning. Principles or evidence might serve as the cause of something or as a final cause; alternatively, principles may serve as moral, juridical, or scientific law.

Principles as Cause

Principles can function in different ways according to the way we express the principle. For the circumstance of cause and effect, principle refers to the cause that was efficient for the effect to come into existence. The principle as cause is traced back to Aristotelian reasoning, which surmises that every event is moved by something prior to it, or has a cause.

Principles as Law

We see principles as law at work in moral law, juridical law, and scientific law. In moral law, principles are what our predecessors teach us as children. "Do unto others," or the "golden rule," is a principle that society has embedded in us so that we are able to function as a civilized people. Principles as moral law are restrictive to the individual as a way of protecting the other person, the whole, or society.

Principles in juridical law are created by the State and also function to limit the liberty of individuals in order to protect the masses. The principles formed in juridical law are written rules that seek to establish a foundation which people can adhere to. The "homestead principle" is an example of a principle in juridical law. The "homestead principle" would function as someone gaining ownership of land because they have made it into a farm, or utilized some resource that has been unused on the land prior to their cultivation of it.

Principles as scientific law function as natural laws, including the Laws of Thermodynamics or natural selection. Principles in scientific law function as laws used to predict certain phenomena that happen in nature. In this context, principles are able to predict results of future experiments. They are developed from facts and also have the ability to be strongly supported by observational evidence.

Determining Whether Evidence is Relevant and Sufficient

Selecting the most relevant material to support a written text is a necessity in producing quality writing and for the credibility of an author. Arguments lacking in reasons or examples won't work in persuading the audience later on, because their hearts have not been pulled. Using examples to support ideas also gives the writing rhetorical effects such as pathos (emotion), logos (logic), or ethos (credibility), all three of which are necessary for a successful text.

An author needs to think about the audience. Are they indifferent? An author might use a personal story or example in order stir empathy. Are they resistant? If so, an author might use logical reasoning based in factual evidence so they will be convinced. Personal stories or testimonials, statistics, or documentary evidence are various types of examples that one can use for their writing.

Determining Whether a Statement Is or Is Not Supported

Evidence used in arguments must be credible and valid, such as that from peer-reviewed scholarly journals. Peer-reviewed sources are sources that have been reviewed by other experts in the field. must also be relevant by being up-to-date, especially those within the science or technology fields.

For example, let's a passage is discussion pesticides and the collapse of bees. An argument without relevant examples would look like this:

> With the use of the world's most popular pesticides, bees are becoming extinct. This is also causing ecological devastation. We must do something soon about the bee population, or else we will chase bees to extinction and lose valuable resources as a result.

Here is the same argument with examples. The added examples are in italics:

> With the use of the world's most popular pesticides, bees are becoming extinct. *Beekeepers have reported losing 55 to 95 percent of their colony in just two short years due to toxic poisoning.* This is also causing ecological devastation. *Bees are known for pollinating more than two-thirds of the world's most essential crops.* We must do something soon about the bee population, or else we will chase bees to extinction and lose valuable resources as a result.

Adding examples to the above argument brings life to the bees—they are living, dying, pollinating—and readers feel more compelled to act as a result of adding relevant examples to the argument.

Assessing Whether an Argument is Valid

Not all arguments are valid. Authors sometimes have one or more flaws in their argument's reasoning. Some SSAT questions may ask you to provide a description of that error. In order for you to be able to describe what flaw is occurring in the argument, it will help to know of various argumentative flaws, such as red herring, false choice, and correlation vs. causation.

Here are some examples of what this type of question looks like:

- The reasoning in the argument is flawed because the argument . . .
- The argument is most vulnerable to criticism on the grounds that it . . .
- Which one of the following is an error in the argument's reasoning?
- A flaw in the reasoning of the argument is that . . .
- Which one of the following most accurately describes X's criticism of the argument made by Y?

Bait/Switch

One common flaw that is good to know is called the "bait and switch." It occurs when the test makers will provide an argument that offers evidence about X, and ends the argument with a conclusion about Y. A "bait and switch" answer choice will look like this:

> The argument assumes that X does in fact address Y without providing justification.

Let's look at an example:

> Hannah will most likely always work out and maintain a healthy physique. After all, Hannah's IQ is extremely high.

The correct answer will look like this:

> The argument assumes that Hannah's high IQ addresses her likelihood of always working out without providing justification.

Ascriptive Error

The ascriptive argument will begin the argument with something a third party has claimed. Usually, it will be something very general, like "Some people say that . . ." or "Generally, it has been said . . ." Then, the arguer will follow up that claim with a refutation or opposing view. The problem here is that when the arguer phrases something in this general sense without a credible source, their refutation of that evidence doesn't really matter. Here's an example:

> It has been said that peppermint oil has been proven to relieve stomach issues and, in some cases, prevent cancer. I can attest to the relief in stomach issues; however, there is just not enough evidence to prove whether or not peppermint oil has the ability to prevent any kind of cancerous cells from forming in the body.

The correct answer will look like this:

> The argument assumes that the refuting evidence matters to the position that is being challenged.

> We have no credible source in this argument, so the refutation is senseless.

Prescriptive Error

First, let's take a look at what "prescriptive" means. Prescriptive means to give directions, or to say something *ought to* or *should* do something else. Sometimes an argument will be a descriptive premise (simply describing) that leads to a prescriptive conclusion, which makes for a very weak argument. This is like saying "There is a hurricane coming; therefore, we should leave the state." Even though this seems like common sense, the logical soundness of this argument is missing. A valid argument is when the truth of the premise leads absolutely to the truth of the conclusion. It's when the conclusion *is* something, not when the conclusion *should* be or do something. The flaw here is the assumption that the conclusion is going to work out; something prescriptive is not ever guaranteed to work out in a logical argument.

False Choice

A false choice, or false dilemma, flaw is a statement that assumes only the object it lists in the statement is the solution, or the only options that exist, for that problem. Here is an example:

I didn't get the grade I want in Chemistry class. I must either be really stupid, I didn't get enough sleep, or I didn't eat enough that day.

This is a false choice error. We are offered only three options for why the speaker did not get the grade he or she wanted in Chemistry class. However, there is potentially more options why the grade was not achieved other than the three listed. The speaker could have been fighting a cold, or the professor may not have taught the material in a comprehensive way. It is our job as test takers to recognize that there are more options other than the choices we are given, although it appears that the only three choices are listed in the example.

Red Herring

A red herring is a point offered in an argument that is only meant to distract or mislead. A red herring will throw something out after the argument that is unrelated to the argument, although it still commands attention, thus taking attention away from the relevant issue. The following is an example of a red herring fallacy:

Kirby: It seems like therapy is moving toward a more holistic model rather than something prescriptive, where the space between a therapist and client is seen more organic rather than a controlled space. This helps empower the client to reach their own conclusions about what should be done rather than having someone tell them what to do.

Barlock: What's the point of therapy anyway? It seems like "talking out" problems with a stranger is a waste of time and always has been. Is it even successful as a profession?

We see Kirby present an argument about the route therapy is taking toward the future. Instead of responding to the argument by presenting their own side regarding where therapy is headed, Barlock questions the overall point of therapy. Barlock throws out a red herring here: Kirby cannot proceed with the argument because now Kirby must defend the existence of therapy instead of its future.

Correlation Versus Causality

Test takers should be careful when reviewing causal conclusions because the reasoning is often flawed, incorrectly classifying correlation as causality. Two events that may or may not be associated with one another are said to be linked such that one was the cause or reason for the other, which is considered the effect. To be a true "cause-and-effect" relationship, one factor or event must occur first (the cause) and be the sole reason (unless others are also listed) that the other occurred (the effect). The cause serves as the initiator of the relationship between the two events.

For example, consider the following argument:

Last weekend, the local bakery ran out of blueberry muffins and some customers had to select something else instead. This week, the bakery's sales have fallen. Therefore, the blueberry muffin shortage last weekend resulted in fewer sales this week.

In this argument, the author states that the decline in sales this week (the effect) was caused by the shortage of blueberry muffins last weekend. However, there are other viable alternate causes for the decline in sales this week besides the blueberry muffin shortage. Perhaps it is summer and many normal patrons are away this week on vacation, or maybe another local bakery just opened or is running a special sale this week. There might be a large construction project or road work in town near the bakery, deterring customers from navigating the detours or busy roads. It is entirely possible that the decline this week is just a random coincidence and not attributable to any factor other than chance, and that next week, sales will return to normal or even exceed typical sales. Insufficient evidence exists to confidently assert that the blueberry muffin shortage was the sole reason for the decline in sales, thus mistaking correlation for causation.

Identifying Assumptions in an Argument and Determining if they are Supported by the Evidence

In the structure of an argument, an **assumption** is an unstated premise. To identify the assumption within arguments, you must find something that the argument is relying on that the author is not stating explicitly. Many strengthening and weakening questions deal with unstated assumptions, as well as necessary and sufficient assumption questions. Let's take a look at what an unsated assumption looks like:

All restaurants in the Seattle area serve vegan food. *Haile's Seafood* must serve vegan food.

Let's identify all parts of the argument, including the unstated assumption. The conclusion of this argument is the last sentence: *Haile's Seafood* must serve vegan food. The premise we are given is the first sentence: All restaurants in the Seattle area serve vegan food. Now let's ask ourselves if there's a missing link. How did the author reach this conclusion? The author reached this conclusion with an unstated assumption, which might look like this: *Haile's Seafood* is in the Seattle area. Now we have the argument:

Premise: All restaurants in the Seattle area serve vegan food.

Unstated Assumption: *Haile's Seafood* is in the Seattle area.

Conclusion: *Haile's Seafood* must serve vegan food

Another way to look at the missing link is like this: there is a connection between "Seattle area" and "vegan food," and one between *"Haile's Seafood"* and "vegan food", but there is no connection between "Seattle area" and *"Haile's Seafood."* The unstated assumption identifies this connection.

Analyzing Two Arguments and Evaluating the Types of Evidence Used to Support Each Claim

Authors use arguments to persuade the reader to agree with their claim. When analyzing the strength of an argument, the reader should summarize the author's message and determine whether or not they agree or disagree with the claim and whether, overall, they believe the evidence the author presented to support it. Next the reader should identify each individual detail and examine it to determine if it supports the claim, fails to support the claim, or even distract from the author's goal.

As readers examine each piece of evidence, they should consider its type and its effectiveness in proving the author's claim both individually and in context. Readers should pay attention to details that contradict the evidence, and readers should question the way the author uses details.

Facts can often be interpreted in ways that mislead the reader, so they will believe a claim that is unproven or untrue. If the claim is supported by facts such as statistics or empirical evidence, the reader should examine them. The reader should question whether the facts are relevant or, if they were interpreted differently, they would lead to a different conclusion that would fail to support the argument or even contradict it. They should consider whether the scope of the data is appropriate, or if it is too broad or narrow, and they should question whether the source is credible, and whether a source has been cited at all.

If the author uses expert opinions, the reader should consider whether the expert is credible and appropriate. If a doctor offers an opinion about pediatric medicine, the reader should ask if the doctor is an expert in that field or if they practice a different form of medicine.

Finally, the reader should consider whether the author has anticipated and adequately responded to any potential counterarguments to the claim and decide if the argument overall is strong, or if the evidence fails to persuade them to accept the author's claim.

Science

Claims and Evidence in Science

Finding Evidence that Supports a Finding

Science passages can contain a lot of information about a specific subject. Generally, there is one main idea and then supporting details. The main idea is the message that the whole passage is conveying, often stated in a complete sentence. The topic of the passage is described in just one word or a short phrase but also conveys a message about the passage. The supporting details provide facts and evidence to support the claim of the main idea. Consider the following passage about earthquakes:

> Earthquakes can be described as either a surface earthquake or a deep-focus earthquake. Surface earthquakes occur when the Earth's crust cracks to relieve stress, a fact agreed upon by most scientists. They occur at depths less than 70 km. The origin of deep-focus earthquakes is debated among scientists. However, most scientists believe they are caused by Earth's tectonic plates sliding toward each other to release pressure from fluids inside the tectonic plates. They occur much deeper than surface earthquakes, at depths of 300 to 700 km below the Earth's surface.

The main idea of this passage is about earthquakes and how they are classified into two categories. The supporting details tell us how the earthquakes are classified into each category. The surface earthquakes occur within 70 km of the Earth's surface and happen as a result of a fracture in the Earth's crust. Deep-focus earthquakes occur between 300 and 700 km below the Earth's surface and happen as a result of Earth's tectonic plates sliding toward each other.

Making Sense of Information that differs Between Various Science Sources

Various scientific sources can present information in different ways. It is important to be able to draw conclusions using information from different sources, such as journal articles written by different scientists on the same topic, to better understand the topic being discussed. For example, when discussing the breeding of flowering plants, one scientist may examine color and the other may investigate genotype. Both sets of research will reveal something about future generations of the plants but they use different methods to look at different aspects of genetics.

Looking at this example in more detail, Scientist A decides to crossbreed one white flowering plant with one red flowering plant. The result is one red flowering plant. She repeats this crossbreeding experiment three more times and gets three more red flowering plants. Scientist B analyzes the genotypes of the parent generation red and white flowering plants and finds that both have homozygous genotypes for the color of their flowers. Putting this information together, it is clear that the red flowers have a dominant phenotype based on the results of the crossbreeding experiment. However, the plants that resulted from the crossbreeding will all have a heterozygous genotype.

Science Vocabulary, Terms, and Phrases

Understanding and Explaining Information from Passages

Scientific information can be presented in many different ways. It can be described using words in a passage, by summarizing or paraphrasing the area of research. Summaries involve putting the main idea of a passage into your own words. Paraphrasing involves condensing the original passage without changing the wording too much. The paraphrased version must be attributed to the original source because the words are not newly generated. It can also be described using graphs, charts, and tables. Chemical reactions can be described using equations and formulas. Putting together information from all of these different formats can create a better understanding of the scientific information being portrayed.

For example, the methods of a scientific experiment can be described in a written passage: Scientist A wants to find out if using a fertilizer containing nitrogen improves plant growth, as nitrogen can help plants produce chlorophyll and other proteins. She plants seeds of the same plant into six pots with soil. She adds fertilizer containing nitrogen to three of the plants and fertilizer without nitrogen to the other three plants. She waters each plant daily and records how tall the plants grow over the course of one month. The results of this experiment can be described in a table with exact numbers:

	Inches	
Day	Average growth for plants getting fertilizer with nitrogen	Average growth for plants getting fertilizer without nitrogen
0	0	0
7	1	0
14	2	0.5
21	3	0.75
28	5	1

The data can also be portrayed using a graph for a more visual representation of the difference in plant growth. The conclusion that the plant receiving fertilizer with nitrogen grew much faster can be drawn immediately from looking at this graph:

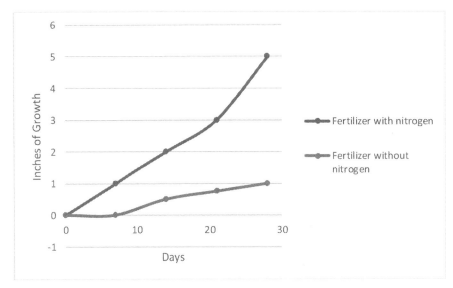

Understanding Symbols, Terms, and Phrases in Science
There are many different symbols, terms, and phrases used when presenting scientific information. When doing a scientific experiment, it is important to first develop a hypothesis about the phenomenon you are

observing and the question you are asking. A **hypothesis** is an educated guess about the answer to the question. It is always a declarative statement about the phenomenon, such as "An area with full sunlight will cause the plant to grow faster than an area without sunlight." The experiment is the methods that are used to test whether or not the hypothesis is true. The phenomenon being observed is the dependent variable in the experiment. The outside factor that may be affecting the phenomenon is the independent variable. Data include all of the measurements that are taken during the experimental process. The data collected should be relevant to the phenomenon being observed and the question being asked. Once the data are collected and analyzed, a conclusion can be drawn. The conclusion will either support or disprove the hypothesis.

Chemical reactions are used to describe many different phenomena. They have special symbols in them and are written as follows:

$$C + O_2 \rightarrow CO_2$$

This reaction represents what happens when carbon is burned, which is the addition of oxygen to carbon. The "+" symbol separates two different molecules, either as reactants or products. The "$\rightarrow$" is a yield symbol that shows the direction of the reaction and separates the reactants from the products. The reactants are written on the left side of the yield symbol and the products are written on the right side of the yield symbol.

Different areas of science also have specific, specialized terms that are used for that subspecialty. It is useful to use context clues within a passage or through the graphic elements to decipher the meaning of these terms. The three main branches of science are physical science, Earth science, and life science. The tables below contain some of the major terms used in each branch.

Physical Science Terms	
Term	**Definition**
Aqueous	A solution made with water
Condensation	The changing of a substance from a gas to a liquid
Element	A substance in which all of the atoms in the sample are alike
Evaporation	The changing of a substance from a liquid to a gas
Half Life	The amount of time it takes for one half of a radioactive isotope to decay
Polymer	A large molecule made up of many smaller molecules linked together
Sublimation	The changing of a substance from a solid to a gas without forming a liquid

Earth Science Terms	
Term	**Definition**
Ablation	The loss of ice or snow from a glacier due to melting, evaporation, or erosion
Delta	A flat, low landform near the mouth of a river
Earthquake	The shaking of the ground caused by a sudden movement of the Earth's crust
Erosion	The process of land being worn down by water or wind
Fossil	The preserved remains or traces of a living organism from the past
Mantle	The layer of earth between the outer crust and the core of the Earth
Topography	A description of the physical features of land

Life Science Terms	
Term	**Definition**
Abiotic	The nonliving features of the environment, including air, water, and soil
Antibody	A protein made to attack an antigen
Antigen	A foreign substance in the body
Cell	The smallest unit of a living thing that can perform the functions of life
DNA	The genetic material of all organisms
Homeostasis	The regulation of an organism's internal environment
Nucleus	The organelle of a cell that controls the cell's activities and contains its genetic information
Tissue	A group of similar cells that work together to perform one job

Using Scientific Words to Express Science Information

It is important to use scientific words to express scientific information so that the information is portrayed as technical, accurate, and useful, as opposed to speculation. Scientific law is different than scientific theory. Scientific law is a phenomenon that has been observed repeatedly. Scientific theory is the explanation or reason behind an observed phenomenon. Hypotheses are used to prove whether or not a theory is true. When collecting and discussing data, measurements should be exact and include units so that comparisons are easy to make. A portion of a whole population or other thing being studied is a sample. Samples are representative of the whole population. Trends are patterns or tendencies that are observed within the data. Predictions are claims or conclusions that can be made based on the trends that are seen. They are a projection of whether or not the hypothesis will be proven to be true. Statistics is a subspecialty of mathematics that helps organize and analyze data to see if differences truly exist between the items that are being compared in the experiment.

Science Investigations

Designing a Science Investigation

When reading scientific passages about experiments, readers should note whether the investigators properly adhered to the scientific method, which is a systemized process for conducting experimental research. The following are crucial steps in this process:

1. Make an observation about a phenomenon.

2. Perform a literature review. This involves researching studies that have already been done and finding out what their results were. This can help form the hypothesis for your scientific investigation.

3. State a clear hypothesis, which is a declarative sentence about what you think the results will show at the end of the experiment.

4. Set up the experimental design. This should include a detailed description of how the data will be collected. There should be a control group, which does not include the variable being tested, in addition to the experimental groups, which have only one variable being changed between them. The design should also include a description about the tools that are being used to make the measurements required. The tools should be capable of making quantitative measurements so that objective comparisons can be made between the groups. Qualitative data does not involve numerical data and can be subjective according to the person making the observations. The objects involved in the experiment should be randomized between all of the groups to remove bias and ensure equal probability of each object receiving a particular treatment. The design should also include a repetition of collecting the measurements to improve accuracy of the treatment results for the groups.

5. After the experiment is completed, the data should be interpreted, which may include the use of statistics to determine the relevance of differences seen between the experimental groups.

6. The investigation should end with a finite conclusion that either supports or disproves the hypothesis.

Examining Hypotheses

A **hypothesis** is a well-defined research statement. An experiment then follows, usually using quantitative research. Quantitative research is research based on empirical data.

The results are then analyzed to determine whether the hypothesis was proven or disproven. Examining a hypothesis is also called **hypothesis testing**. Examining a hypothesis happens most often in science, and it isn't really appropriate for social sciences such as social studies and history. However, qualitative hypotheses can be made in these disciplines to further examine a social or historical event. A hypothesis, in this light, should clearly state the argument that the writer wishes to examine, and the reason or reasons why the author feels it is relevant. This type of hypothesis statement generally requires the "what" and the "why." Consider the following qualitative hypothesis:

"The Métis in Canada were less discriminated against than were Canada's First Nations since they were partly descendants of European fur traders."

The first half of the hypothesis—"The Métis in Canada were less discriminated against than were Canada's First Nations"—reveals the "what," and the second part—"since they were partly descendants of European fur traders"—is the "why."

In science, hypotheses are generally written as "if, then" statements that require the collection of unbiased, empirical, and quantitative data to either prove or disprove the hypothesis. Consider the following:

"**If** a hibiscus flower is placed in direct sunlight and watered twice a day, **then** it will thrive."

In order for conclusions to be accepted as valid and credible, it is extremely important that the data collected isn't biased. The researchers must consider all possible angles of the study or experiment, and they must refrain from collecting the data in such a way as to purposely prove the hypothesis. Conducting studies and experiments of this nature helps to advance the different disciplines, challenge widely accepted beliefs, and broaden a global understanding of the fields of social studies, history, and the sciences.

Identifying Possible Errors in a Science Investigation and Changing the Design to Correct Them

Scientific investigations are subject to either random or systematic errors. **Random errors** are the differences in data measurements created by the precision limitations of the instrument being used to take the measurements. It is nearly impossible for the investigator to take the same measurement in the exact same way repeatedly throughout an experiment. There will be random, minute fluctuations with every measurement.

For example, a cell biologist measuring the volume of media in a petri dish uses a pipet to make her measurement. She takes three measurements, and they range between 4.95 ml and 5.05 ml. Similarly, a chemist takes the weight of a sodium chloride sample three times using the same scale and finds the weights to be 14.2 g, 14.1 g, and 14.4 g. This type of error can be minimized or eliminated by taking more data and increasing the number of observations. Then, a statistical evaluation can determine whether or not the differences in the measurements are meaningful or simply random error.

Systematic errors are fluctuations in the measurements in only one direction and are reproducible. They are caused by a flaw in the device used for measurement or by an error made consistently by the investigator. These types of errors cannot be determined by statistical analysis. Instead, the experimental design and procedure must be closely examined to determine where the error is being made. For example, the temperature of a chemical solution is measured by an electronic thermometer and the measurements are consistently low because the thermometer has not been calibrated. Another example is that the length of a fossil is consistently off by an inch in all measurements because the investigator is taking the measurement starting at the one-inch mark instead of the zero mark.

Identifying the Strengths and Weaknesses of Different Types of Science Investigations

There are several factors that determine how strong or weak a scientific investigation is. Participants or items must be assigned to a group randomly. There should not be a bias in determining which participants or items are placed in which group. Control groups must be included to account for the many variables that are not being tested by the experiment. The control group should be treated identically to the experimental group but without receiving the experimental treatment. Only one variable should be tested in the experiment, while all other variables are consistent between all of the groups. The results that are collected must be quantifiable in order to be comparable between the groups.

Three types of scientific investigations are descriptive, comparative, and experimental. A **descriptive investigation** includes an experimental design, data collection and analysis, and a conclusion. A **comparative investigation** compares two or more things against each other. It does not include a control group to account for more than the variable being tested. It includes a hypothesis, an experimental design that includes independent and dependent variables, data collection and analysis, and a conclusion. An **experimental investigation** includes a hypothesis, an experimental design that includes independent and dependent variables, control and experimental groups, data collection and analysis, and a conclusion, and is the most thorough and strongest type of investigation.

Using Evidence to Draw Conclusions or Make Predictions

Deciding Whether Conclusions are Supported by Data

Once the results of an investigation are analyzed, a conclusion can be drawn based on the evidence collected. The conclusion is limited in scope to the items or participants tested in the investigation. If an unbiased, representative sample was tested, the conclusion can be broadened to the whole population but not beyond that. To determine whether a conclusion is supported by the data, logic and reasoning skills have to be used to analyze the data and evidence collected in the investigation.

In an investigation determining whether pineapple trees can survive in Florida, the conclusion was made that pineapple trees cannot survive in Florida. Data were collected by recording the amount of rainfall in Florida for three months. Pineapple trees need wet climates to survive and produce pineapples, with an average requirement of five inches of water each month. The data shows that there was one inch of water in May, two inches of water in June, and one inch of water in July. Because pineapple trees need five inches of water per month, the conclusion of the investigation is supported by the data collected, which detail the amount of water that is actually available for pineapple trees in Florida.

Making Conclusions Based on Data

Data can either be **qualitative** (observations, interviews, or focus groups) or **quantitative** (measured data, as in the population of a certain country, a person's height, or the depth of the Earth's oceans). Once fully understood and analyzed, the data can now be interpreted. For instance, how has the analysis of this data affected the students' initial assumptions, thoughts, and beliefs? Data interpretation helps to make the knowledge gleaned from a study more meaningful.

Analysis of data must precede interpretation. Data analysis can take many forms. For instance, various relationships within the data can be identified. Certain patterns or trends may come to light that help to better grasp the meaning or the relevance of the results. The opposite could also be true; a thorough analysis of data may yield that there are not any patterns, trends, or relationships. After fully analyzing the data, the information can start to be interpreted, which helps the reader of the study make informed decisions. Interpreting the data means acquiring a greater understanding of the results of the data. For example, the analysis of collected data on temperature patterns throughout the year and around the globe might reveal that there is a pattern of increasingly hotter temperatures. The interpretation of this data may then result in an individual's greater understanding of global warming.

Making Predictions Based on Data

Predictions can have a greater scope than the conclusions of the experiment. They can use the data collected along with limited conclusions that were made to apply a conclusion to a larger population or similar populations that were not tested in the original investigation. Predictions can include a hypothesis that predicts the outcome if the same experiment was completed on a different set of items or a different population.

Let's say an investigation is done to look at hair growth with a certain vitamin in yellow Labrador retrievers. It was concluded that the vitamin does not help with hair growth. The scientist could then predict how the same vitamin would affect chocolate Labradors or other breeds of similar dogs. Similarly, thinking about the pineapple trees again, if they do not grow well in Florida because of their need for an abundance of rainfall every month, the scientist could predict how other tropical trees would grow in Florida. Knowing that banana plants require a similar amount of rainfall, it could be predicted that they would not grow well in Florida. However, orange trees only require one inch of water a month, so it could be predicted that they would grow well in Florida.

Science Theories and Processes

Science theories are explanations about different aspects of the natural world. They are based on observations and experiments that have been done to obtain evidence about a certain phenomenon, so they are not random guesses. **Scientific processes** are used to investigate the theories and determine whether the theories hold true or false. The scientific process starts with asking a question, formulating a hypothesis about the question, establishing a method for collecting data, and then collecting and analyzing the data. All of these parts put together help to answer the original question and test the scientific theory.

There have been many scientific theories developed in history. In the 1770s, Antoine Lavoisier proposed the oxygen theory of combustion and figured out that oxygen was the element that was combining with other substances to make them burn, which opposed the theory that every combustible substance contained phlogiston (a fire-like element) within it. Einstein developed the theory of special relativity in 1905 and found that space and time are interwoven into one continuum, known as space-time. Thus, two events could happen at the same time for one observer but at different times for another observer.

A more tangible example is that a group of doctors wants to determine whether or not there are more complications with women who deliver their babies at home compared with women who deliver their babies in the hospital. They believe that there are more complications with delivery at home, which would be the scientific theory. The doctors may have formulated this theory based on the patients they have seen and information they received from other patients and doctors, but they have not investigated it themselves, and so do not have any of their own evidence or data. Once the theory is developed, they need to follow an appropriate scientific process to answer their question.

They choose two hundred patients at random to enter their investigation after making sure the patients all have similar healthy pregnancies. Half of the patients have chosen to deliver their babies at home and half have chosen to deliver at the hospital. They count the number of complications that occur with the deliveries both at home and in the hospital. Comparing these numbers, the doctors can see which number is higher and where more complications happen. Ten of the women who delivered at home had complications, whereas five of the women who delivered at the hospital had complications. The doctors conclude that there is a greater chance of complications with deliveries that occur at home, which answers their question and proves their scientific theory to be true.

Science Formulas and Statistics

Applying Science Formulas

There are many standard formulas used to make calculations on scientific data. They can be used on many different sets and types of data to determine the same outcome measure. It is important to critically read the passage to figure out what information is given and what information is missing and being asked for. This will allow you to figure out what formula needs to be applied. For example, science utilizes three primary temperature scales. The temperature scale most often used in the United States is the Fahrenheit (F) scale. The Fahrenheit scale uses key markers based on the measurements of the freezing (32 °F) and boiling (212 °F) points of water. In the United States, when taking a person's temperature with a thermometer, the Fahrenheit scale is used to represent this information. The human body registers an average temperature of 98.6 °F.

Another temperature scale commonly used in science is the Celsius (C) scale (also called **centigrade** because the overall scale is divided into one hundred parts). The Celsius scale marks the temperature for water freezing at 0 °C and boiling at 100 °C. The average temperature of the human body registers at 37 °C. Most countries in the world use the Celsius scale for everyday temperature measurements.

For scientists to easily communicate information regarding temperature, an overall standard temperature scale was agreed upon. This scale is the Kelvin (K) scale. Named for Lord Kelvin, who conducted research in thermodynamics, the Kelvin scale contains the largest range of temperatures to facilitate any possible readings.

The Kelvin scale is the accepted measurement by the International System of Units (from the French *Système international d'unités*), or SI, for temperature. The Kelvin scale is employed in thermodynamics, and its reading for 0 is the basis for absolute zero. This scale is rarely used for measuring temperatures in the medical field.

The conversions between the temperature scales are as follows:

Degrees Fahrenheit to Degrees Celsius:

$$^0C = \frac{5}{9}(^0F - 32)$$

Degrees Celsius to Degrees Fahrenheit:

$$^0F = \frac{9}{5}(^\circ C) + 32$$

Degrees Celsius to Kelvin:

$$K = \,^0C + 273.15$$

For example, if a patient has a temperature of 38 °C, what would this be on the Fahrenheit scale?

Solution:

First, select the correct conversion equation from the list above.

$$^0F = \frac{9}{5}(^\circ C) + 32$$

Next, plug in the known value for °C, 38.

$$^0F = \frac{9}{5}(38) + 32$$

Finally, calculate the desired value for °F.

$$^0F = \frac{9}{5}(38) + 32$$

$$^\circ F = 100.4 ^\circ F$$

For example, what would the temperature 52 °C be on the Kelvin scale?

First, select the correct conversion equation from the list above.

$$K = \,^0C + 273.15$$

Next, plug in the known value for °C, 52.

$$K = 52 + 273.15$$

Finally, calculate the desired value for K.

$$K = 325.15 \, K$$

Using Statistics to Describe Science Data

Statistics involves the analysis of quantitative data to determine something about a whole population using a representative sample. It not only includes calculations of the numbers collected, but also considers how the numbers were collected and other important factors about the experimental design. There are many different statistical tests that can be used depending on how the investigation was set up and the data were collected. Statistical tests help to interpret scientific data, which helps determine whether or not the original question was answered and if the hypothesis holds true or not.

The term "significant" is often used to explain the results of a statistical test. If the test shows that the differences seen between the experimental groups are consistent enough to also apply to the larger population, the difference is described as significant. Sometimes differences are seen between experimental groups but they are too small or not consistent enough to hold true through the statistical test. In this case, the investigator would not be able to say that the differences seen in the experimental sample can also be seen in the population as a whole.

Probability and Sampling in Science

Determining the Probability or Likelihood of Something Happening

Probability is the chance, or likelihood, of an event occurring. With enough information, the probability of any event can be calculated. The likelihood of one single event happening is called **simple probability** and can be calculated as follows:

Simple probability = (# of favorable outcomes) / (# of total possible outcomes)

Favorable outcomes are the events that you want to determine the probability of. Probability can be expressed as a decimal, fraction, or using the term "___ out of ___." For example, let's say you want to determine the probability of rolling a number 4 on a die. There is only one number 4 out of six total numbers. You can divide one (the number of 4s on the die) by six (the total number of numbers on the die) and find that the probability of rolling a number 4 on a die is 0.17 or 1/6 or 1 out of 6.

Compound probability is more complex than simple probability and would allow you to calculate the probability of rolling a number 4 on a die if you rolled more than one time. It determines the likelihood of multiple events happening and is calculated by multiplying the probabilities of each event happening by each other.

Compound probability = (Probability of event #1) x (Probability of event #2) x and so on...

Using a Sample to Answer Science Questions

Often when trying to answer a scientific question, it is not possible to collect data from a whole population or all items in question. Scientific investigations use a sample of the population to collect evidence and data and then draw a conclusion that can be applied to the whole population. The sample size must be large enough to give a true representation of the population and is based on the acceptable margin of error and the confidence level, both of which are statistical calculations. **Stratified sampling** can also help create a representative sample population. The whole population is divided into several homogeneous subpopulations and then a portion of each subpopulation is chosen to be represented in the sample population.

Here is an example of the use of sampling to determine a characteristic about the whole population. Scientist A wants to determine how many high school students in Rochester, NY carry backpacks that weigh more than ten pounds. It would take a lot of time to weigh the backpacks of the thousands of high school students in the city and would also be a lot of data to analyze. So, Scientist A decides to take a sample of the high school student population and weigh only their backpacks to answer her question. She randomly chooses 250 students from high schools throughout the city to participate in her study and weighs their backpacks. She finds that two hundred of her sample students have backpacks that weigh more than ten pounds. Applying her conclusion to the whole population, she can say that eighty percent of the high school students in Rochester, NY carry backpacks that weigh more than ten pounds.

Using Counting to Solve Science Problems

Many scientific investigations involve the quantification of a sample. In order to quantify the portion of a sample that has a certain feature, that portion must be counted. Comparing different portions of the sample with each other can reveal different facets of the population or sample being studied. The objects of a study can be arranged according to shape, size, color, or any other feature, and the feature chosen will affect the proportion of the sample that is counted. **Permutations** allow for the calculation of the number of different ways a certain number of objects can be arranged. For example, if there are four seats a table, the number of permutations tells us how many different seating arrangements are possible to fill the seats with four different people. Where n represents the number of positions to fill, there would be n choices of people for the first seat, $(n-1)$ choices for the second seat, $(n-2)$ choices for the third seat, and 1 choice for the fourth seat. Permutations are calculated by multiplying these together, giving us the following formula:

$$_nP_n = n \times (n-1) \times (n-2) \times (n-3) \times 1$$

Plugging in $n = 4$ for this case, there are 24 different possible seating arrangements.

Presenting Science Information Using Numbers, Symbols, and Graphics

Using Graphics to Display Science Information

Graphics are a useful tool for presenting data in a visual, rather than a descriptive, manner. Tables, charts, and graphs are some of the ways that data can be presented through graphics. These tools often make it easier to make comparisons between different experimental groups and to visualize when large changes are happening during the experimental process.

For a table, the experimental groups are listed in the header or first row, and the independent variable values are listed in the first column. The rest of the table is filled in with data collected about the dependent variable. Here is an example of a table with data about bacterial growth in the presence of different antibiotics:

	Number of Bacteria (thousands)		
	Day 0	**Day 7**	**Day 14**
Control	0	10	40
Antibiotic A	0	2	4
Antibiotic B	0	5	15

A bar graph can be used to display the same data:

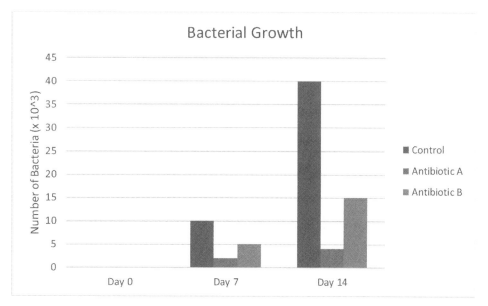

A line graph can also be used to show the same data:

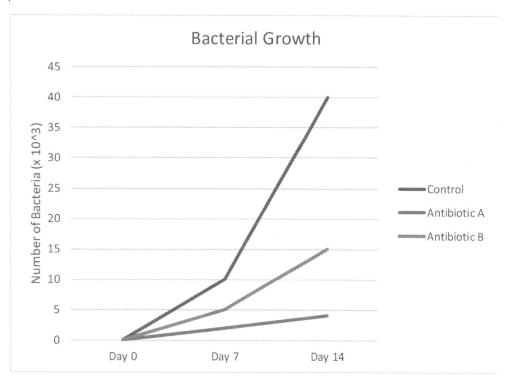

Pie charts are a useful tool for representing different portions of a population:

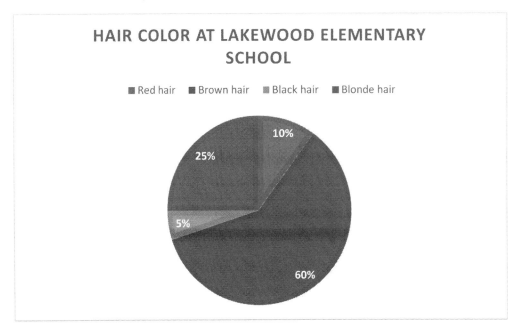

HAIR COLOR AT LAKEWOOD ELEMENTARY SCHOOL

■ Red hair   ■ Brown hair   ■ Black hair   ■ Blonde hair

Each of these graphics displays the same information but allows the data to be visualized in different ways.

Scientific Notation

Scientific notation is a system used to represent numbers that are very large or very small. Sometimes, numbers are way too big or small to be written out with multiple zeros behind them or in decimal form, so scientific notation is used as a way to express these numbers in a simpler way.

Scientific notation takes the decimal notation and turns it into scientific notation, like the table below:

Decimal Notation	Scientific Notation
5	5×10^0
500	5×10^2
10,000,000	1×10^7
8,000,000,000	8×10^9
-55,000	-5.5×10^4
.00001	10^{-5}

In scientific notation, the decimal is placed after the first digit and all the remaining numbers are dropped. For example, 5 becomes "5.0×10^0." This equation is raised to the zero power because there are no zeros behind the number "5." Always put the decimal after the first number. Let's say we have the number 125,000. We would write this using scientific notation as follows: 1.25×10^5, because to move the decimal from behind "1" to behind "125,000" takes five counts, so we put the exponent "5" behind the "10." As you can see in the table above, the number ".00001" is too cumbersome to be written out each time for an equation, so we would want to say that it is "10^{-5}." If we count from the place behind the decimal point to the number "1," we see that we go backwards 5 places. Thus, the "-5" in the scientific notation form represents 5 places to the right of the decimal.

Social Studies

Main Ideas and Details in Social Studies Readings

Determining the Main Ideas

Determining the main idea of a social studies text is much like determining the main idea of any other text. The **main idea** is the author's central message; it is the chief argument that pervades the *entire* text. The main idea of each paragraph in the text is typically different from the overall main idea. However, the main ideas of the paragraphs typically reinforce the main idea of the entire text. When searching for the main idea of a social studies text, one must first ask whether the main idea is being conveyed explicitly or implicitly. **Explicit main ideas** are directly expressed to the audience. **Implicit main ideas** are implied by the language and literary devices of the text. Explicit main ideas are likely to be found either at the beginning of the text (typically within the first sentence of the introductory paragraph) or in the thesis of the text (which, more often than not, can be found near the conclusion of the introductory paragraph). Most historical court cases, for example, have explicit main ideas—their rulings are delivered in direct language. When the main idea is not explicitly stated, it is more difficult to identify. Implicit main ideas must be deduced or decoded by gathering together the facts, arguments, and images hinted at in the text and drawing information from these clues. Some historical speeches, poems, songs, and political cartoons have implicit main ideas. They are likely to incorporate indirect literary devices to express their main points.

Whether the text has an implicit or explicit main idea, every student should begin with one overarching question: "What point is the author trying to make in this text?"

This overarching question should be followed by several smaller questions:

1. **Who**?—Who wrote the text and who (if anybody) is the text describing?

2. **What**?—What time frame is being discussed by the text? What historical context is it implicitly representative of? What context is it explicitly discussing?

3. **Where**?—Where was the text written? Which region, culture, or place is its main focus?

4. **When**?—When was the text written? Is the date explicitly referenced? Or is it implicitly implied?

5. **Why**?—Why did the author write this text? Is there a good reason or explanation for its existence?

6. **How**?—How is the theory or method of interpretation presented to the audience? Likewise, how did the text even originate?

If you can successfully answer the overarching question and the majority of the underlying questions, then you have successfully pinpointed the main idea of a text.

Now, let's cross-analyze a text with an explicit main idea and a text with an implicit main idea.

DOCUMENT A: EXPLICIT MAIN IDEA SOCIAL STUDIES TEXT

"(2) No person acting under color of law shall—"(A) in determining whether any individual is qualified under State law or laws to vote in any Federal election, apply any standard, practice, or procedure different from the standards, practices, or procedures applied under such law or laws to other individuals within the same county, parish, or similar political subdivision who have been found by State officials to be qualified to vote;

"(B) deny the right of any individual to vote in any Federal election because of an error or omission on any record or paper relating to any application, registration, or other act requisite to voting, if such error or omission is not material in determining whether such individual is qualified under State law to vote in such election."

Voting Rights Act of 1965

DOCUMENT B: IMPLICIT MAIN IDEA SOCIAL STUDIES TEXT

I speak tonight for the dignity of man and the destiny of Democracy. I urge every member of both parties, Americans of all religions and of all colors, from every section of this country, to join me in that cause.

At times, history and fate meet at a single time in a single place to shape a turning point in man's unending search for freedom. So it was at Lexington and Concord. So it was a century ago at Appomattox. So it was last week in Selma, Alabama. There, long-suffering men and women peacefully protested the denial of their rights as Americans. Many of them were brutally assaulted. One good man—a man of God—was killed.

There is no cause for pride in what has happened in Selma. There is no cause for self-satisfaction in the long denial of equal rights of millions of Americans. But there is cause for hope and for faith in our Democracy in what is happening here tonight. For the cries of pain and the hymns and protests of oppressed people have summoned into convocation all the majesty of this great government—the government of the greatest nation on earth. Our mission is at once the oldest and the most basic of this country—to right wrong, to do justice, to serve man. In our time we have come to live with the moments of great crises. Our lives have been marked with debate about great issues, issues of war and peace, issues of prosperity and depression.

But rarely in any time does an issue lay bare the secret heart of America itself. Rarely are we met with a challenge, not to our growth or abundance, or our welfare or our security, but rather to the values and the purposes and the meaning of our beloved nation. The issue of equal rights for American Negroes is such an issue. And should we defeat every enemy, and should we double our wealth and conquer the stars, and still be unequal to this issue, then we will have failed as a people and as a nation. For, with a country as with a person, "what is a man profited if he shall gain the whole world, and lose his own soul?"

There is no Negro problem. There is no Southern problem. There is no Northern problem. There is only an American problem.

And we are met here tonight as Americans—not as Democrats or Republicans; we're met here as Americans to solve that problem. This was the first nation in the history of the world to be founded with a purpose.

The great phrases of that purpose still sound in every American heart, North and South: "All men are created equal." "Government by consent of the governed." "Give me liberty or give me death." And those are not just clever words, and those are not just empty theories. In their name Americans have fought and died for two centuries and tonight around the world they stand there as guardians of our liberty risking their lives. Those words are promised to every citizen that he shall share in the dignity of man. This dignity cannot be found in a man's possessions. It cannot be found in his power or in his position. It really rests on his right to be treated as a

Both documents discuss American citizens' response to the civil rights movement in different ways. Document A is and excerpt from the Voting Rights Act of 1965. It discusses an explicit legal response to the abrogation of citizens' rights to vote. This act makes legal provisions that protect people's voting rights. Document B is an excerpt from a famous speech made by President Lyndon B. Johnson before Voting Rights Act of 1965 was ratified. President Johnson makes an implicit references to the civil rights movement in the United States by referring to Selma, Alabama, site of a key civil rights march. He ties the civil rights movement to other key points in US history (Lexington and Concord, as well as Appomattox). In referring to these events, he is implicitly tying Dr. Martin Luther King's civil rights marches in Alabama to the broader search for freedom and the longer march for democracy in US history. In doing this, he is implicitly claiming racism and the voting disenfranchisement of African-Americans are barriers to democracy. He is also implicitly stating that the division between North and South is one that must end. Together, these texts show how main ideas can be either explicit (the Voting Rights Act of 1965) or implicit (the "We Shall Overcome" Speech).

Using Details to Make Inferences or Claims

Whether their main idea is explicit or implicit, all texts employ details that enable the audience to make inferences. Let's once again use the example of Lyndon B. Johnson's "We Shall Overcome" speech to gain a better understanding of how details build up claims.

Take a look at some of the details highlighted in bold below.

DOCUMENT C: EXAMPLES OF DETAILS IN TEXTS

(1) **I speak tonight for the dignity of man and the destiny of Democracy**. I urge every member of both parties, Americans of all religions and of all colors, from every section of this country, to join me in that cause.

(2) **At times, history and fate meet at a single time in a single place to shape a turning point in man's unending search for freedom. So it was at Lexington and Concord. So it was a century ago at Appomattox. So it was last week in Selma, Alabama.** There, long-suffering men and women peacefully protested the denial of their rights as Americans. **(3) Many of them were brutally assaulted. One good man—a man of God—was killed.**

There is no cause for pride in what has happened in Selma. There is no cause for self-satisfaction in the long denial of equal rights of millions of Americans. But there is cause for hope and for faith in our Democracy in what is happening here tonight. For the cries of pain and the hymns and protests of oppressed people have summoned into convocation all the majesty of this great government—the government of the greatest nation on earth. Our mission is at once the oldest and the most basic of this country—to right wrong, to do justice, to serve man. In our time we have come to live with the moments of great crises. **(4) Our lives have been marked with debate about great issues, issues of war and peace, issues of prosperity and depression.**

(5) **But rarely in any time does an issue lay bare the secret heart of America itself.** Rarely are we met with a challenge, not to our growth or abundance, or our welfare or our security, but rather to the values and the purposes and the meaning of our beloved nation. The issue of equal rights for American Negroes is such an issue. And should we defeat every enemy, and should we double our wealth and conquer the stars, and still be

unequal to this issue, then we will have failed as a people and as a nation. For, with a country as with a person, **(6) "But what is a man profited if he shall gain the whole world, and lose his own soul?"**

(7) There is no Negro problem. There is no Southern problem. There is no Northern problem. There is only an American problem.

And we are met here tonight as Americans—not as Democrats or Republicans; we're met here as Americans to solve that problem. **(8) This was the first nation in the history of the world to be founded with a purpose.**

(9) The great phrases of that purpose still sound in every American heart, North and South: "All men are created equal." "Government by consent of the governed." "Give me liberty or give me death." And those are not just clever words, and those are not just empty theories. In their name Americans have fought and died for two centuries and tonight around the world they stand there as guardians of our liberty risking their lives. Those words are promised to every citizen that he shall share in the dignity of man. This dignity cannot be found in a man's possessions. It cannot be found in his power or in his position. It really rests on his right to be treated as a man equal in opportunity to all others. It says that he shall share in freedom. He shall choose his leaders, educate his children, provide for his family according to his ability and his merits as a human being.

President Lyndon B. Johnson. "We Shall Overcome" Speech (March 15, 1965)

These details form the philosophical fabric of the excerpt; they combine to form evidence that can be used to make claims. Details in texts can take the form of facts, opinions, references or citations, or literary devices. The sentences and phrases in bold above are examples of these kinds of details. The first bold sentence—*I speak tonight for the dignity of man and the destiny of Democracy*—emphasizes American democracy, tying it to the dignity of humanity. The phrase partially personifies democracy by providing it with something typically afforded to people: a destiny.

In the second bold section—*At times, history and fate meet at a single time in a single place to shape a turning point in man's unending search for freedom. So it was at Lexington and Concord. So it was a century ago at Appomattox. So it was last week in Selma, Alabama*—the author is using another literary device to provide details: allusion. President Johnson is using allusions to important historical events that championed freedom and unity in the United States to support his claim that Selma is yet another wave of increasing, broadening democracy.

At times, Johnson steps away from the literary devices in order to cite the facts of the matter, as he does in the third bold section, where he reports that "Many of [the protesters at Selma] were brutally assaulted. One good man—a man of God—was killed."

Allusion, which seems to be the kind of detail most used in this speech, also emerges in the fourth bold section, where Johnson discusses debates over abolition, wars (which includes the Civil War over slavery and abolition, among other wars), and the Great Depression. These allusions place history on the side of the president and on the side of the Selma protesters.

The fifth bold section once again employs personification, humanizing America by giving it a beating heart, one that is plagued by racial hatred and injustice. In order to challenge the ethics of the audience,

Johnson even cites the Bible in the sixth bold section: "But what is a man profited if he shall gain the whole world, and lose his own soul?" This is a quotation from Matthew 16:26 in the New Testament, which contrasts the emptiness of material gain with the fullness of spiritual peace.

Sometimes authors use syntactical techniques such as repetition to make their claims known. Johnson does so in this passage: "There is no Negro problem. There is no Southern problem. There is no Northern problem. There is only an American problem." The repetition of the phrase "there is no [x]problem" emphasizes the problem is shared by all Americans rather than limited to one segment or region.

To bolster his arguments, Johnson sometimes uses hyperbolic statements like the one presented in the eighth bold section: "This was the first nation in the history of the world to be founded with a purpose."

The final bold section returns to allusion, noting the historical tensions between North and South, and then it concludes with two citations, one from the Declaration of Independence and one on from Patrick Henry's famous "Give Me Liberty or Give Me Death" revolutionary speech. These are just a few examples of the ways in which in-text details can be used to make inferences and claims.

The biggest take-away, however, is that these details are essentially useless without historical background knowledge. Background knowledge allows us to make more logical inferences.

How Authors Use Language in Social Studies

When a reader is evaluating the point of view or purpose of a historical text or piece of art, he or she must be aware of how language is used to convey the author's message(s). Authors and cartoonists or painters can incorporate concrete facts or strategic imagery into a text or work of art to help establish their point of view or purpose. Language exists in both text-based forms and visual art forms. Language can be used to convey facts, opinions, references or citations, or literary devices. The example of Dr. Martin Luther King's "I Have a Dream" Speech has already helped exhibit the power, complexity, and utility of language. There are countless social studies sources that employ the power of language. Consider this 1962 speech by John F. Kennedy as another example:

> ... **(1) William Bradford, speaking in 1630 of the founding of the Plymouth Bay Colony, said that all great and honorable actions are accompanied with great difficulties, and both must be enterprised and overcome with answerable courage.**
>
> If this **(2) capsule history of our progress** teaches us anything, it is that man, in his quest for knowledge and progress, is determined and cannot be deterred. The exploration of space will go ahead, whether we join in it or not, and it is one of the great adventures of all time, and no nation which expects to be the leader of other nations can expect to stay behind in the race for space.
>
> Those who came before us made certain that this country **(3) rode the first waves of the industrial revolutions, the first waves of modern invention, and the first wave of nuclear power, and this generation does not intend to founder in the backwash of the coming age of space.** We mean to be a part of it—we mean to lead it. For the eyes of the world now look into space, to the moon and to the planets beyond, and we have vowed that we shall not see it governed by a hostile flag of conquest, but by a banner of freedom and peace. **(4) We have vowed that we shall not see space filled with weapons of mass destruction, but with instruments of knowledge and understanding.**
>
> Yet the vows of this Nation can only be fulfilled if we in this Nation are first, and, therefore, we intend to be first. In short, our leadership in science and in industry, our hopes for peace and security, our obligations to

> ourselves as well as others, all require us to make this effort, to solve these mysteries, to solve them for the good of all men, and to become the world's leading space-faring nation....
>
> ...(5) **We choose to go to the moon. We choose to go to the moon in this decade and do the other things,** not because they are easy, but because they are hard, because that goal will serve to organize and measure the best of our energies and skills, because that challenge is one that we are willing to accept, one we are unwilling to postpone, and one which we intend to win, and the others, too...
>
> *President John F. Kennedy, "We Choose to Go to the Moon" Speech, September 12, 1962*

Readers should notice how John F. Kennedy (JFK) uses language—including facts, opinions, references, and literary devices such as imagery—to establish his point of view and purpose. In the first highlighted passage JFK uses a historical reference; he refers to the Plymouth Colony in order to connect the early history of the United States to what he hopes will be its future: flights to the moon. He uses this reference to historical fact to stir the emotions of the audience and to highlight Americans' destiny as courageous explorers.

In the second highlighted passage, JFK uses the phrase "capsule history of progress" to help the audience recall the history he has just summarized. In the third highlighted passage, JFK uses the image of progress acting like a wave; this is a play on words, a double meaning. The waves refer to eras of sweeping change in history, but also actual waves, which conjure up notions of seafaring discovery. When he says that "this nation does not intend to founder in the backwash of the coming space age," he is again employing aquatic imagery, encouraging people to move forward instead of falling back. The fourth highlighted passage sets forth an opinion: JFK declares that he does not want weapons of mass destruction (a reference to the Cold War) planted in space. Finally, the repetition of "we choose to go to the moon" in the final paragraph drives the point home, making it a collective endeavor rather than an individual one.

Language—in the form of facts, opinions, references, and literary devices—is used in this speech to make one major point: JFK believes America should be the first nation to land on the moon.

Fact versus Opinion

A **fact** is a statement that is true empirically or an event that has actually occurred in reality, and can be proven or supported by evidence; it is generally objective. In contrast, an **opinion** is subjective, representing something that someone believes rather than something that exists in the absolute. People's individual understandings, feelings, and perspectives contribute to variations in opinion. Though facts are typically objective in nature, in some instances, a statement of fact may be both factual and yet also subjective. For example, emotions are individual subjective experiences. If an individual says that they feel happy or sad, the feeling is subjective, but the statement is factual; hence, it is a subjective fact. In contrast, if one person tells another that the other is feeling happy or sad—whether this is true or not— that is an assumption or an opinion.

Claims and Evidence in Social Studies

Determining Whether a Claim Is or Is Not Supported by Evidence
There are three major ways to determine whether a claim in social studies is or is not supported by evidence: 1) cross-referencing the claim with information provided in-text, 2) cross-referencing the claim with information provided in *other* texts, 3) cross-referencing the claim with information generally accepted as fact, according to our background knowledge.

For example, someone could write:

Coach Smith's basketball team did not improve during his years as head coach.

There are three ways you can evaluate this statement.

1. Cross-referencing the claim with information provided in-text. The writer could then list the win/loss record of the team during his time as coach:

Year	Coach	Wins	Losses	Winning Percentage
2015	Taylor	14	6	70%
2016	Smith	13	7	65%
2017	Smith	11	9	55%
2018	Smith	11	9	55%

The 2015 season is the year before Smith became the head coach. In his first year (2016) as coach, the team's record decreased to a 65% winning percentage. It decreased again in 2017 to 55% and remained the same in 2018. By providing information in the article, you can see if the data supports the writer's claim. Of course, the writer could have made up the information, so you must evaluate if the source seems trustworthy. You can also verify the data by using a secondary source.

2. Cross-referencing the claim with information provided in *other* texts. To do this, you must find another article, book, or source for your information. Perhaps the wins and losses of the basketball teams are recorded on the basketball league's website or in other news articles.

3. Cross-referencing the claim with information generally accepted as fact, according to our background knowledge. In other cases, you may follow the basketball team so closely that you know from memory that this is true.

Sometimes you know from your own sources and knowledge that something is correct or incorrect. For example, if someone said that the U.S. Senate had 100 members, you may not have to verify it if you remember from school or previous learning that it has 100 members.

Comparing Information that Differs Between Sources
A **primary source** is a piece of original work. This can include books, musical compositions, recordings, movies, works of visual art (paintings, drawings, photographs), jewelry, pottery, clothing, furniture, and other artifacts. Within books, primary sources may be of any genre. Whether nonfiction based on actual events or a fictional creation, the primary source relates the author's firsthand view of some specific event, phenomenon, character, place, process, ideas, field of study or discipline, or other subject matter. Whereas primary sources are original treatments of their subjects, **secondary sources** are a step removed from the original subjects; they analyze and interpret primary sources. These include journal articles, newspaper or magazine articles, works of literary criticism, political commentaries, and academic textbooks.

In the field of history, primary sources frequently include documents that were created around the same time period that they were describing, and most often produced by someone who had direct experience or knowledge of the subject matter. In contrast, secondary sources present the ideas and viewpoints of other authors about the primary sources; in history, for example, these can include books and other written works about the particular historical periods or eras in which the primary sources were produced. Primary sources pertinent in history include diaries, letters, statistics, government information, and original journal articles and books. In literature, a primary source might be a literary novel, a poem or book of poems, or a play. Secondary sources addressing primary sources may be criticism, dissertations, theses, and journal articles. **Tertiary sources,** typically reference works referring to primary and secondary sources, include encyclopedias, bibliographies, handbooks, abstracts, and periodical indexes.

In scientific fields, when scientists conduct laboratory experiments to answer specific research questions and test hypotheses, lab reports and reports of research results constitute examples of primary sources. When researchers produce statistics to support or refute hypotheses, those statistics are primary sources. When a scientist is studying some subject longitudinally or conducting a case study, they may keep a journal or diary. For example, Charles Darwin kept diaries of extensive notes on his studies during sea voyages on the *Beagle*, visits to the Galápagos Islands, etc.; Jean Piaget kept journals of observational notes for case studies of children's learning behaviors. Many scientists, particularly in past centuries, shared and discussed discoveries, questions, and ideas with colleagues through letters, which also constitute primary sources. When a scientist seeks to replicate another's experiment, the reported results, analysis, and commentary on the original work is a secondary source, as is a student's dissertation if it analyzes or discusses others' work rather than reporting original research or ideas.

Making Inferences

Essentially, inferences are educated guesses based on the presented evidence and the reader's background knowledge that help form conclusions in the absence of one being directly stated. Inferences should be conclusions based off of sound evidence and reasoning.

Connections Between Different Social Studies Elements

Describing the Connections Between People, Places, Environments, Processes, and Events
History is different than mathematics and hard sciences because it deals with human phenomena that are difficult to define: people, places, environments, processes, and events. The best way to define these phenomena is to find connections, which are typically extracted from historical sources—which can be both primary and secondary sources—and artifacts. It may be helpful to think of historians as "forensic detectives" who must find clues by analyzing evidence from the past. These clues help to create conceptual connections that show how certain people, places, environments, processes, and events are connected.

Let's take a look at the Industrial Revolution to understand how multiple factors can fit together:

> During the Industrial Revolution, factories moved from using individual labor to machines and assembly lines. As part of this chance, industry began using fossil fuels such as coal and oil over energy sources like water and wood. The industry changes began in Great Britain, and then much of the innovation spread to the United States. The ability to accomplish and create more generated economic growth. This increased prosperity made way for greater food, more stable lives, and health innovations that led to a drop in infant mortality rates and increases in life expectancies. Both of those factors led to extreme population rise.

The chart below shows how these factors affected one another:

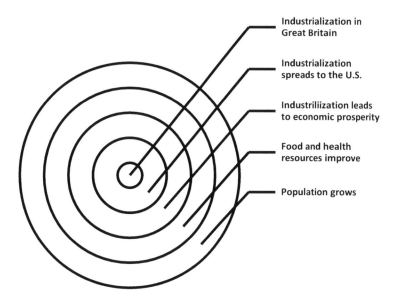

Industrialization in
Great Britain

Industrialization
spreads to the U.S.

Industriliization leads
to economic prosperity

Food and health
resources improve

Population grows

Putting Events in Order and Understanding the Steps in a Process
One of the biggest tasks for any historian or student is placing events in order so that the steps of the historical process can be understood. Historical events are linked by causation (cause and effect) and correlation (associations that are not cause and effect). In order to gain a better understanding of the bigger historical picture, it helps to think about the order in which things happen. In some cases, historians might take a step-by-step look at one event's effects over time; in other cases, historians might try to list several events within one era.

Take a look at the two examples that follow:

EXAMPLE A: ONE EVENT'S EFFECTS OVER TIME: Watergate Scandal

Year	Date	Event
1972	June 17	Watergate Break-in
1973	May 17	Senate hearings begin
	June 25	John Dean begins testifying
	October 20	Saturday Night Massacre (string of events reflecting poorly on Nixon)
1974	July 27	Judiciary Committee votes for impeachment of Nixon
	August 5	Release of "Smoking Gun" tape
	August 8	Nixon announces his resignation
	August 9	Nixon resigns
	August 9	Ford inaugurated as President
	September 8	Ford pardons Nixon
	October 17	Ford testifies before Congress regarding the pardon

European Colonization

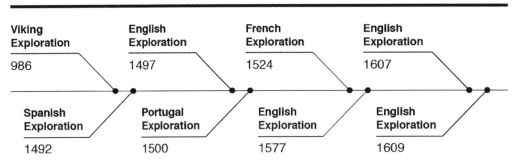

Notice how both timelines place events in order, according to date, and evaluate their impact on history. However, the first timeline looks at the impact of one event—the Watergate Scandal—on a broad span of history, while the second timeline looks at many events and categorizes them under one broad theme—European colonization. Yet both timelines are great examples of how historians place events in order, according to a process, which in this case, is time.

Analyzing the Relationship of Events, Processes, and/or Ideas
As social science students and scholars create timelines or conceptual maps of events, processes, identities, or ideas, they can evaluate whether earlier events caused later events, simply occurred before

these events, or indirectly affected these events. In order to do this, they must contextualize or historicize the events, processes, identities, and ideas they encounter in sources. Likewise, they must compare these sources to other sources—via cross-textual analysis or cross-referencing—in order to validate their relationships between events, processes, identities, and ideas.

Here are some questions that a person can ask in order to illuminate these events:

- Which event, process, or idea came first?

- Are the events, processes, and ideas causally or correlatively connected?

- If the events, processes, or ideas are causally related, then how significant influence did the cause have on the effect?

- If the events, processes, or ideas are correlatively related, then how direct or indirect is the relationship?

- Is there anything that may prove this causal or correlative relation wrong?

- What evidence do you have to support or prove all of the above answers?

Let's attempt this vetting process for analyzing the relationships of two events, processes, or ideas. For this example, let's try to analyze the relationship between the Japanese attack on Pearl Harbor and the United States' entry into World War II.

- Which event, process, or idea came first?
  ANSWER: The attack on Pearl Harbor happened before the United States' entry into World War II.

- Are the events, processes, and ideas causally or correlatively connected?
  ANSWER: These events are causally correlated because one led to another.

- If the events, processes, or ideas are causally related, then how significantly did the preceding event influence the later one?
  ANSWER: There is an extremely significant influence of the attack on the declaration of war.

- If the events, processes, or ideas are correlatively related, then how direct or indirect is the relationship?
  ANSWER: The events are causally related, not correlatively related, so this question is not applicable.

- Is there anything that may prove this causal or correlative relation wrong?
  ANSWER: Some may argue that the entry of the US into WWII was inevitable, but this is a weaker argument in comparison to the more generally accepted causal argument. This argument is speculative in comparison to the causal argument.

- What evidence do you have to support or prove all of the above answers?
  ANSWER: Perhaps the best evidence is President Franklin D. Roosevelt's speech to Congress requesting a declaration of war. See the evidence below for confirmation.

Mr. Vice President, Mr. Speaker, members of the Senate and the House of Representatives:

Yesterday, December 7th, 1941—a date which will live in infamy—the United States of America was suddenly and deliberately attacked by naval and air forces of the Empire of Japan.

The United States was at peace with that nation, and, at the solicitation of Japan, was still in conversation with its government and its Emperor looking toward the maintenance of peace in the Pacific.

Indeed, one hour after Japanese air squadrons had commenced bombing in the American island of Oahu, the Japanese Ambassador to the United States and his colleague delivered to our Secretary of State a formal reply to a recent American message. And, while this reply stated that it seemed useless to continue the existing diplomatic negotiations, it contained no threat or hint of war or of armed attack.

It will be recorded that the distance of Hawaii from Japan makes it obvious that the attack was deliberately planned many days or even weeks ago. During the intervening time the Japanese Government has deliberately sought to deceive the United States by false statements and expressions of hope for continued peace.

The attack yesterday on the Hawaiian Islands has caused severe damage to American naval and military forces. I regret to tell you that very many American lives have been lost. In addition, American ships have been reported torpedoed on the high seas between San Francisco and Honolulu.

Yesterday the Japanese Government also launched an attack against Malaya.
Last night Japanese forces attacked Hong Kong.
Last night Japanese forces attacked Guam.
Last night Japanese forces attacked the Philippine Islands.
Last night the Japanese attacked Wake Island.
And this morning the Japanese attacked Midway Island.

Japan has therefore undertaken a surprise offensive extending throughout the Pacific area. The facts of yesterday and today speak for themselves. The people of the United States have already formed their opinions and well understand the implications to the very life and safety of our nation.

As Commander-in-Chief of the Army and Navy I have directed that all measures be taken for our defense, that always will our whole nation remember the character of the onslaught against us.

No matter how long it may take us to overcome this premeditated invasion, the American people, in their righteous might, will win through to absolute victory.

I believe that I interpret the will of the Congress and of the people when I assert that we will not only defend ourselves to the uttermost but will make it very certain that this form of treachery shall never again endanger us.

Hostilities exist. There is no blinking at the fact that our people, our territory and our interests are in grave danger.

With confidence in our armed forces, with the unbounding determination of our people, we will gain the inevitable triumph. So help us God.

I ask that the Congress declare that since the unprovoked and dastardly attack by Japan on Sunday, December 7th, 1941, a state of war has existed between the United States and the Japanese Empire.

-President Franklin D. Roosevelt—December 8, 1941

Cross-referencing evidence is crucial to ensuring that the perceived connections do, indeed, exist. In this particular instance, the evidence supporting a causal relationship is very clear: "I ask that the Congress declare that since the unprovoked and dastardly attack by Japan on Sunday, December 7th, 1941, a state of war has existed between the United States and the Japanese Empire." There are, however, times when the causal relation might not be as clear or the evidence might not be as firm. This forces historians to comb through thousands of documents to find the right clues to validate perceived connections. Regardless, this process of questioning can be used to vet and validate relationships.

The Effect of Different Social Studies Concepts on an Argument or Point Of View

Analyzing How Events and Situations Shape the Author's Point of View

There are many things we human beings can overcome; history is not one of these things. We can make history, shape history, and analyze history. We can even write history or change our views of history. But we cannot undo what happened in the past. Put simply, *our historical context inevitably shapes our points of view*. We cannot escape—sociologically speaking—many of the events and situations that shape our character. Thus, when you are analyzing primary and secondary sources in social studies, keep in mind that the authors of these sources are always influenced, some more than others, by their historical context.

Thus, whenever we consider an author's point of view, we must also ask the following questions to help us historicize or contextualize their opinions:

Can we validate the exact date or era of the source?

Certain events and certain eras live within particular paradigms in history. A paradigm is worldview or overarching pattern of thought that underlies the theories and methodology of a particular field of knowledge, in this case, history. These paradigms help us understand people's ideas and people's actions.

Are there any direct or indirect references to other historical events, moments, ideas, or figures?

If you do not know the exact date or era of the source, you may be able to approximate it by identifying direct or indirect references to other historical events, moments, ideas, or figures.

Are there any implicit or explicit citations in the source that hint at the intellectual lineage of the author?

Most ideas or points of view do not emerge in a historical vacuum. It is human inclination to pilfer ideas from others. We all owe much of our intellect to specific books, authors, or mentors. Sometimes these books, authors, and mentors are alluded to in a text or piece of art.

If there is not a reference to time, is there a reference to place or culture?

We are all shaped by nationalism, regionalism, localism, and culturalism. References to place or culture can provide clues to how events and situations shape an author's point of view.

Evaluating Whether the Author's Evidence is Factual, Relevant, and Sufficient

There are three major ways students and scholars can evaluate whether an author's evidence is factual, relevant, and sufficient.

1. **Background Knowledge and Fact-Checking**: In order to evaluate an author's evidence, every student or scholar must have a broad background knowledge. In social studies, students or scholars must be able to reference events, facts, primary sources, secondary sources, theories, and historiographies. This background knowledge allows students and scholars to fact-check particular statements. Much like a journalist fact-checks the statements of their interviewees, a student or scholar of social studies must fact-check the statements of a source.

2. **Cross-Referencing Sources**: Sometimes the truth becomes muddled in history. One event can have millions of sources, which begs the question "What is Truth?" The best way to figure out the truth is to cross-reference and triangulate sources, paying close attention to common statements, phrases, perspectives, and themes.

3. **Deconstructing Biases**: Every source has biases that need to be deconstructed. These biases can either hide the truth, and, in some cases, they illuminate the truth. For examples, biases may help historians historicize or contextualize a certain source because every context or era has its own set of paradigmatic biases.

Making Judgments About How Different Ideas Impact the Author's Argument
In order to make judgments about how different ideas impact the author's argument, you must first follow the same pattern as evaluating sources. However, there are some slight nuances to this approach and one additional step in the process.

1. **Background Knowledge and Fact-Checking:** In order to evaluate an author's arguments, every student must some background knowledge. In social studies, students must be able to reference events, facts, primary sources, secondary sources, theories, and historiographies. This background knowledge allows students to fact-check particular statements. Much like a journalist fact-checks the statements of their interviewees, a student of social studies must fact-check the statements of a source. This process allows you to validate an author's claims.

2. **Cross-Referencing Sources:** Who else is making these arguments? Do you trust the person or people making these arguments? Are they a reliable source? Do the author's arguments have some semblance of collective validity'? In order to validate an argument, it helps to either a) discover the source of the argument, or b) understand the academic lineage of an argument. This can be accomplished through cross-referencing sources.

3. **Examining Biases:** While every argument has biases, the best arguments are as objective as possible. How objective or subjective is the argument? The only way you can answer this question is by examining the biases of an argument. The biases can be illuminated through background knowledge, fact-checking, and cross-referencing.

4. **Comparing the Arguments to Your Own Worldview:** We all bring our own biases, worldviews, theories, and arguments to the sources we encounter. The sources students and scholars use are filtered through their own perspectives. Thus, since this process is about making judgments rather than simply evaluating, we must be prepared to become aware of our own biases. That is the only way we can address the biases we encounter in other sources.

Identifying Bias and Propaganda in Social Studies Readings

All sources have a relative amount of bias. However, when the bias of the historical materials verges on coercion, censorship, or indoctrination, then the source becomes what we call **propaganda.** Propaganda is different than bias. Bias can be conscious or unconscious. But propaganda is always conscious and strategic. Propaganda can best be described as an attempt at swaying the beliefs or opinions of a target audience. Some people may refer to propaganda as brainwashing. The best way to identify propaganda is to: a) be aware of what is fact and fiction in history (background knowledge), b) be aware of the ways in which truth can be manipulated for certain causes, and c) be conscious of intensified historical prejudices.

Biases

Biases usually occur when someone allows their personal preferences or ideologies to interfere with what should be an objective decision. In personal situations, someone is biased towards someone if they favor them in an unfair way. In academic writing, being biased in your sources means leaving out objective information that would turn the argument one way or the other. The evidence of bias in academic writing makes the text less credible, so be sure to present all viewpoints when writing, not just your own, so to avoid coming off as biased. Being objective when presenting information or dealing with people usually allows the person to gain more credibility.

Stereotypes

Stereotypes are preconceived notions that place a particular rule or characteristics on an entire group of people. Stereotypes are usually offensive to the group they refer to or allies of that group, and often have negative connotations. The reinforcement of stereotypes isn't always obvious. Sometimes stereotypes can be very subtle and are still widely used in order for people to understand categories within the world. For example, saying that women are more emotional and intuitive than men is a stereotype, although this is still an assumption used by many in order to understand the differences between one another.

Using Data Presented in Visual Form, Including Maps, Charts, Graphs, and Tables

Making Sense of Information that is Presented in Different Ways

To make arguments, social studies fields—history, political science, sociology, geography, philosophy, law, and criminal justice—make use of both qualitative data (words, quotes, passages) and quantitative data (numbers and statistics). This qualitative and quantitative information can be presented in a variety of ways. In most cases, the qualitative and quantitative information will be highlighted in in-text references or quotations. Take a look at the two examples below to gain a better understanding of what these in-text references look like.

EXAMPLE A: QUALITATIVE INFORMATION REFERENCE

According to [Émile] Durkheim, such a society produces, in many of its members, **psychological states characterized by a sense of purposelessness, emotional emptiness and despair.**

EXAMPLE B: QUANTITATIVE INFORMATION REFERENCE

According to a report released by the Milwaukee Police Department in 2015, **the median homicide rate per 100,000 residents was 23, but District 5 had a murder rate of 55.**

Analyzing Information from Maps, Photographs, and Political Cartoons

Besides in-text citations, data and information can also be presented in the forms of maps and graphs.

Take a look at how information is represented spatially on the map below by showing the state lines and the areas of each land purchases. The information on this map is not only spatially represented, but also color coded through the use of shades of gray.

Photographs and political cartoons are different than maps, tables, and charts because they normally employ aesthetic or artistic aspects that are typically lacking in these quantitative platforms for data visualization. In fact, in most photographs or political cartoons, statistics are altogether absent.

Take a look at the political cartoon below to gain a better understanding of how data or evidence is available even in the absence of statistics.

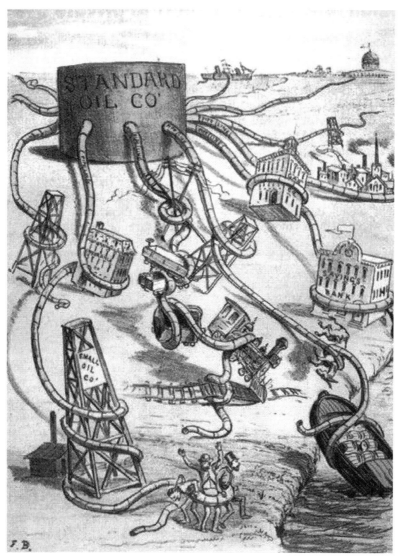

The political cartoon above does not have any numbers (quantitative data) for analysis or interpretation. Nevertheless, it does offer some data. It may even be easier to consider this evidence as historical clues. This particular political cartoon has several clues:

1. An oil storage tank is labeled with the name Standard Oil company. This means that the oil storage tank represents that company.

2. The tank is shown as an octopus causing terror. This means the cartoonist is depicting the oil company in a negative way.

3. The octopuses' limbs are holding onto oil rigs, transportation (ships and railroads), banks, and governmental buildings as if it has control over all of it.

4. One of the oil rigs is labeled "Small Oil Co.", showing that it is stifling out competition.

173

While the clues offer a lot of insight, it is background knowledge that the viewer uses to make inferences about what the image's claims are. Background knowledge of oil companies in the early 1900s and of anti-monopoly laws allows you to use the clues in the political cartoon to understand this cartoon. The cartoon depicts a very large and "evil" Standard Oil Company having control over business, infrastructure, government, and competitors because of its size and power. Public distrust of the size and power of the company eventually led to laws that caused it to break up into smaller companies. Political cartoons, therefore, are best interpreted when one has adequate background knowledge.

Photographs, like political cartoons, offer clues about history. However, they are different from political cartoons because they are somewhat less manipulated than drawings. Political cartoons are created. Photographs are also created, but with less choice; the photographer takes a picture of what shows up in the viewfinder. Both are artifacts of history, but only photographs capture history in "real time."

Take a look at one of these historical snapshots below:

Unlike the political cartoon example, there are no words here, just images. Yet this image is still starkly powerful. The photograph depicts a man standing in front of four tanks. He appears to be confronting the tanks, staring at them directly. The tanks have red stars on them (though these are difficult to see) and they appear to be stopped in their tracks by the lone man.

Without even understanding history, one can appreciate the powerful protest in this picture. However, background knowledge allows us to have greater access to the meaning and significance of the photo. This photo, entitled "Tank Man," was taken in Tiananmen Square in China in 1989. The man is trying to agitate for democratic reform by protesting the brute power of the communist regime (symbolized by the red star and tanks).

Thus, the absence of quantitative data in photographs and political cartoons does not necessarily mean that these important historical artifacts are devoid of data. Yet, much like maps, graphs, and charts, these historical tools are best interpreted when one has adequate background knowledge.

Representing Textual Data into Visual Form

Sometimes it helps to represent textual data in visual form in order to reach a broader audience or reinforce a main argument. Visual images engage more audience members; they also strengthen textual data. That is why many students and scholars in social studies use charts, graphs, maps, and tables to support their claims.

Take the following statement about the annual rainfall in Woodriver as an example:

Rainfall in Woodriver was high in 2018 but not unprecedented.

This in-text statement is sufficient by itself. It is saying that while rainfall was high in 2018, it has been that high before. Nevertheless, this qualitative claim is even more powerful when it is backed by quantitative data in a map, chart, graph, or table.

Take a look at how this type of statement might be repeated or amplified in visual form:

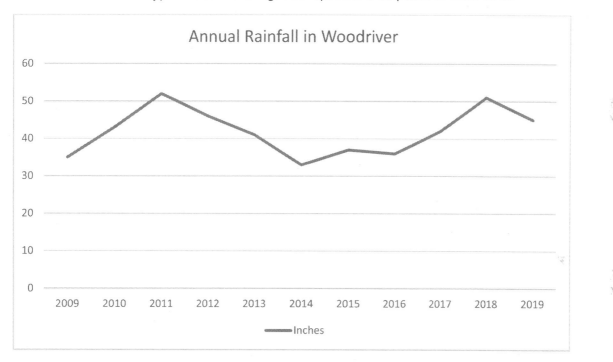

Graphs—or other visual forms like the one above—can help support or reiterate claims by illustrating quantitative examples. Notice how the actual claim—_"Rainfall in Woodriver was high in 2018 but not unprecedented,"_—makes its way into visual form. The graph is accompanied by a title: "Annual Rainfall in Woodriver." The line graph charts rainfall each year in the city of Woodriver from 2009 to 2019, illustrating that, although the rainfall was high in 2018, it was not unprecedented in the city. The year 2011 had even higher rainfall. This visualization of data breathes life into an already powerful statement. In social studies, this type of visualization is often repeated in order to quantify or qualify statements.

Interpreting, Using, and Creating Graphs with Appropriate Labeling, and Using the Data to Predict Trends
All graphs should be interpreted, used, and created with appropriate labels, which include (but are not limited to) the following labels:

The Main Title: The main title offers a brief explanation of what is in the graph. Titles help the audience to understand the main point of a graph.

The Subtitle: The subtitle offers more specific information about the purpose of the graph. Subtitles are brief sentences or phrases that enhance main titles. In some cases, the subtitle can be placed below or beside the main title.

X-Axis and X-Axis Label: Bar graphs and line graphs have an *x*-axis, which runs horizontally (flat). The *x*-axis has quantities representing different categories, statistics, or times that are being compared.

Y-Axis and Y-Axis Label: Bar graphs and line graphs have a *y*-axis, which runs vertically (up and down). The *y*-axis usually measures quantities, typically starting at zero or another designated number.

Take a look at the graph below to get a better understanding of where each label needs to be. This bar graph has the following elements:

>**Title**: What kind of pet do you own?

>**Y-Axis Label**: Number of People

>**X-Axis Label**: Kind of Pet

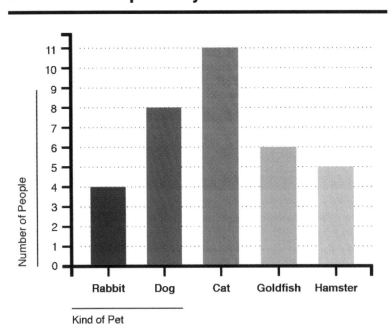

Some graphs also incorporate **keys**, which define colors and symbols. Notice how the graph below uses a key to explain the meaning of each color.

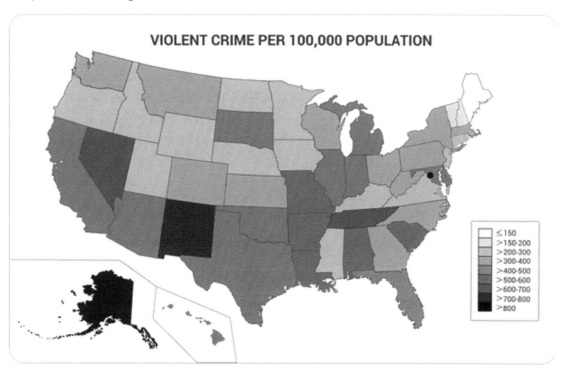

VIOLENT CRIME PER 100,000 POPULATION

≤150
>150-200
>200-300
>300-400
>400-500
>500-600
>600-700
>700-800
>800

Keys, titles, subtitles, and axes are incorporated in order to both clarify arguments, engage a broader audience of learners, and offer alternative forms of organizing data.

Dependent and Independent Variables

In social studies, students and scholars must constantly analyze variables. Variables in social studies refer to people, places, eras, events, things, or phenomena that students and scholars try to measure or evaluate. In essence, students and scholars are trying to evaluate relationships between certain historical people, places, eras, events, things, or phenomena. In history, students or scholars must constantly be asking, "How are these two variables related?" In answering this question, one can determine which variables are dependent on each other and which variables are independent of each other.

Dependent variables literally *depend* on other variables or factors. As a result, they change accordingly as these other variables or factors change. Take the Japanese attack on Pearl Harbor, for example. Most scholars would agree that this attack forced the US to declare its entry into World War II. Thus, the dependent variable in this case would be the United States' entry into World War II; it was a dependent effect or result of the Japanese attack on Pearl Harbor. In fact, most dependent variables are considered to be effects, while most independent variables are considered to be the root causes in social studies. Thus, the independent variable in this case would be the Japanese attack on Pearl Harbor. It is the variable that accounts for the effect on a dependent variable. In this case, the cause-and-effect relationship is easy to decipher; however, this is not always the case when analyzing history and social studies.

Correlation Versus Causation
When analyzing variables in social studies, it is important to distinguish between correlation and causation. Correlation and causation are not always easy to infer in social studies. Correlation does not imply cause-and-effect; rather, it simply implies that there is an implied (typically a *statistically* implied) connection or relationship between two variables. This connection or relationship does not necessarily clearly designate a cause-and-effect relationship. Correlations can be positive or direct, or they can also be negative or indirect. An example of a correlation would be the rise of fascist governments during the Great Depression. The Great Depression did not create fascist governments, but many fascist governments drew inspiration from the economic hardships of the Depression.

In causation, one phenomenon is the direct result of another; this is called a cause-and-effect relationship. A connection can be called **causation** only if three major conditions are met:

- The cause has to precede the effect in time; the cause happens first. For example, the Japanese attack on Pearl Harbor clearly preceded United States' entry into World War II.

- There has to be empirical evidence that supports the claim of causation; there has to be qualitative or quantitative data that supports the cause-and-effect relationship. For instance, references to the Japanese attack in the United States' declaration of war are support the claim of causation.

- The effect cannot be explained away by other variables. For instance, it is difficult to understand causation when it comes to the Great Depression because there are so many variables that contributed to (or are correlated with) the stock market crash in 1929.

Causation, unlike correlation, is concrete: it proves that *this* led to *that*. Correlation, on the other hand, is much less defined. It implies that this *might have* contributed to that.

Using Statistics in Social Studies

Mean, Median, Mode, and Range of a Data Set
In statistics, measures of central tendency are measures of average. They include the mean, median, mode, and midrange of a data set. The **mean**, otherwise known as the **arithmetic average**, is found by dividing the sum of all data entries by the total number of data points. The **median** is the midpoint of the data points. If there is an odd number of data points, the median is the entry in the middle. If there is an even number of data points, the median is the mean of the two entries in the middle. The **mode** is the data point that occurs most often. Finally, the **midrange** is the mean of the lowest and highest data points. Given the spread of the data, each type of measure has pros and cons. In a **right-skewed distribution**, the bulk of the data falls to the left of the mean. In this situation, the mean is on the right of the median and the mode is on the left of the median. In a **normal distribution,** where the data are evenly distributed on both sides of the mean, the mean, median, and mode are very close to one another. In a **left-skewed distribution**, the bulk of the data falls to the right of the mean. The mean is on the left of the median and the mode is on the right of the median.

Here is an example of each type of distribution:

Left Skew

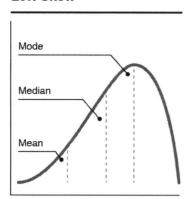

Normal Distribution

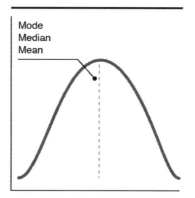

Right Skew

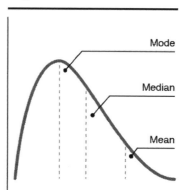

Practice Questions

Questions 1-6 are based on the following passage from The Life, Crime, and Capture of John Wilkes Booth *by George Alfred Townsend:*

The box in which the President sat consisted of two boxes turned into one, the middle partition being removed, as on all occasions when a state party visited the theater. The box was on a level with the dress circle; about twelve feet above the stage. There were two entrances—the door nearest to the wall having been closed and locked; the door nearest the balustrades of the dress circle, and at right angles with it, being open and left open, after the visitors had entered. The interior was carpeted, lined with crimson paper, and furnished with a sofa covered with crimson velvet, three arm chairs similarly covered, and six cane-bottomed chairs. Festoons of flags hung before the front of the box against a background of lace.

President Lincoln took one of the arm-chairs and seated himself in the front of the box, in the angle nearest the audience, where, partially screened from observation, he had the best view of what was transpiring on the stage. Mrs. Lincoln sat next to him, and Miss Harris in the opposite angle nearest the stage. Major Rathbone sat just behind Mrs. Lincoln and Miss Harris. These four were the only persons in the box.

The play proceeded, although "Our American Cousin," without Mr. Sothern, has, since that gentleman's departure from this country, been justly esteemed a very dull affair. The audience at Ford's, including Mrs. Lincoln, seemed to enjoy it very much. The worthy wife of the President leaned forward, her hand upon her husband's knee, watching every scene in the drama with amused attention. Even across the President's face at intervals swept a smile, robbing it of its habitual sadness.

About the beginning of the second act, the mare, standing in the stable in the rear of the theater, was disturbed in the midst of her meal by the entrance of the young man who had quitted her in the afternoon. It is presumed that she was saddled and bridled with exquisite care.

Having completed these preparations, Mr. Booth entered the theater by the stage door; summoned one of the scene shifters, Mr. John Spangler, emerged through the same door with that individual, leaving the door open, and left the mare in his hands to be held until he (Booth) should return. Booth who was even more fashionably and richly dressed than usual, walked thence around to the front of the theater, and went in. Ascending to the dress circle, he stood for a little time gazing around upon the audience and occasionally upon the stage in his usual graceful manner. He was subsequently observed by Mr. Ford, the proprietor of the theater, to be slowly elbowing his way through the crowd that packed the rear of the dress circle toward the right side, at the extremity of which was the box where Mr. and Mrs. Lincoln and their companions were seated. Mr. Ford casually noticed this as a slightly extraordinary symptom of interest on the part of an actor so familiar with the routine of the theater and the play.

1. Which of the following best describes the author's attitude toward the events leading up to the assassination of President Lincoln?
   a. Excitement, due to the setting and its people
   b. Sadness, due to the death of a beloved president
   c. Anger, due to the impending violence
   d. Neutrality, due to the style of the report
   e. Apprehension, due to the crowd and their ignorance

2. What does the author mean by the last sentence in the passage?
   a. Mr. Ford was suspicious of Booth and assumed he was making his way to Mr. Lincoln's box.
   b. Mr. Ford assumed Booth's movement throughout the theater was due to being familiar with the theater.
   c. Mr. Ford thought that Booth was making his way to the theater lounge to find his companions.
   d. Mr. Ford thought that Booth was elbowing his way to the dressing room to get ready for the play.
   e. Mr. Ford thought that Booth was coming down with an illness due to the strange symptoms he displayed.

3. Given the author's description of the play "Our American Cousin," which one of the following is most analogous to Mr. Sothern's departure from the theater?
   a. A ballet dancer who leaves the New York City Ballet just before they go on to their final performance.
   b. A basketball player leaves an NBA team and the next year they make it to the championship but lose.
   c. A lead singer leaves their band to begin a solo career, and the band's sales on their next album drop by 50 percent.
   d. A movie actor who dies in the middle of making a movie and the movie is made anyway by actors who resemble the deceased.
   e. A professor who switches to the top-rated university for their department only to find the university they left behind has surpassed his new department's rating.

4. Which of the following texts most closely relates to the organizational structure of the passage?
   a. A chronological account in a fiction novel of a woman and a man meeting for the first time.
   b. A cause-and-effect text ruminating on the causes of global warming.
   c. An autobiography that begins with the subject's death and culminates in his birth.
   d. A text focusing on finding a solution to the problem of the Higgs boson particle.
   e. A text contrasting the realities of life on Mars versus life on Earth.

5. Which of the following words, if substituted for the word *festoons* in the first paragraph, would LEAST change the meaning of the sentence?
   a. Feathers
   b. Armies
   c. Adornments
   d. Buckets
   e. Boats

6. What is the primary purpose of the passage?
   a. To persuade the audience that John Wilkes Booth killed Abraham Lincoln
   b. To inform the audience of the setting wherein Lincoln was shot
   c. To narrate the bravery of Lincoln and his last days as President
   d. To recount in detail the events that led up to Abraham Lincoln's death
   e. To disprove the popular opinion that John Wilkes Booth is the person who killed Abraham Lincoln

Questions 7-13 are based on the following passage from The Story of Germ Life *by Herbert William Conn:*

The first and most universal change effected in milk is its souring. So universal is this phenomenon that it is generally regarded as an inevitable change that cannot be avoided, and, as already pointed out, has in the past been regarded as a normal property of milk. To-day, however, the phenomenon is well understood. It is due to the action of certain of the milk bacteria upon the milk sugar which converts it into lactic acid, and this acid gives the sour taste and curdles the milk. After this acid is produced in small quantity its presence proves deleterious to the growth of the bacteria, and further bacterial growth is checked. After souring, therefore, the milk for some time does not ordinarily undergo any further changes.

Milk souring has been commonly regarded as a single phenomenon, alike in all cases. When it was first studied by bacteriologists it was thought to be due in all cases to a single species of micro-organism which was discovered to be commonly present and named *Bacillus acidi lactici.* This bacterium has certainly the power of souring milk rapidly, and is found to be very common in dairies in Europe. As soon as bacteriologists turned their attention more closely to the subject it was found that the spontaneous souring of milk was not always caused by the same species of bacterium. Instead of finding this *Bacillus acidi lactici* always present, they found that quite a number of different species of bacteria have the power of souring milk, and are found in different specimens of soured milk. The number of species of bacteria that have been found to sour milk has increased until something over a hundred are known to have this power. These different species do not affect the milk in the same way. All produce some acid, but they differ in the kind and the amount of acid, and especially in the other changes which are effected at the same time that the milk is soured, so that the resulting soured milk is quite variable. In spite of this variety, however, the most recent work tends to show that the majority of cases of spontaneous souring of milk are produced by bacteria which, though somewhat variable, probably constitute a single species, and are identical with the *Bacillus acidi lactici.* This species, found common in the dairies of Europe, according to recent investigations occurs in this country as well. We may say, then, that while there are many species of bacteria infesting the dairy which can sour the milk, there is one that is more common and more universally found than others, and this is the ordinary cause of milk souring.

When we study more carefully the effect upon the milk of the different species of bacteria found in the dairy, we find that there is a great variety of changes they produce when they are allowed to grow in milk. The dairyman experiences many troubles with his milk. It sometimes curdles without becoming acid. Sometimes it becomes bitter, or acquires an unpleasant "tainted" taste, or, again, a "soapy" taste. Occasionally, a dairyman finds his milk becoming slimy, instead of souring and curdling in the normal fashion. At such times, after a number of hours, the milk becomes so slimy that it can be drawn into long threads. Such an infection proves very troublesome, for many a time it persists in spite of all attempts made to remedy it. Again, in other cases the milk will turn blue, acquiring about the time it becomes sour a beautiful sky-blue colour.

Or it may become red, or occasionally yellow. All of these troubles the dairyman owes to the presence in his milk of unusual species of bacteria which grow there abundantly.

7. The word *deleterious* in the first paragraph can be best interpreted as meaning which one of the following?
   a. Amicable
   b. Smoldering
   c. Luminous
   d. Ruinous
   e. Virtuous

8. Which of the following best explains how the passage is organized?
   a. The author begins by presenting the effects of a phenomenon, then explains the process of this phenomenon, and then ends by giving the history of the study of this phenomenon.
   b. The author begins by explaining a process or phenomenon, then gives the history of the study of this phenomenon, this ends by presenting the effects of this phenomenon.
   c. The author begins by giving the history of the study of a certain phenomenon, then explains the process of this phenomenon, then ends by presenting the effects of this phenomenon.
   d. The author begins by giving a broad definition of a subject, then presents more specific cases of the subject, then ends by contrasting two different viewpoints on the subject.
   e. The author begins by contrasting two different viewpoints, then gives a short explanation of a subject, then ends by summarizing what was previously stated in the passage.

9. What is the primary purpose of the passage?
   a. To inform the reader of the phenomenon, investigation, and consequences of milk souring
   b. To persuade the reader that milk souring is due to *Bacillus acidi lactici*, which is commonly found in the dairies of Europe
   c. To describe the accounts and findings of researchers studying the phenomenon of milk souring
   d. To discount the former researchers' opinions on milk souring and bring light to new investigations
   e. To narrate the story of one researcher who discovered the phenomenon of milk souring and its subsequent effects

10. What does the author say about the ordinary cause of milk souring?
   a. Milk souring is caused mostly by a species of bacteria called *Bacillus acidi lactici*, although former research asserted that it was caused by a variety of bacteria.
   b. The ordinary cause of milk souring is unknown to current researchers, although former researchers thought it was due to a species of bacteria called *Bacillus acidi lactici.*
   c. Milk souring is caused mostly by a species of bacteria identical to that of *Bacillus acidi lactici*, although there are a variety of other bacteria that cause milk souring as well.
   d. The ordinary cause of milk souring will sometimes curdle without becoming acidic, though sometimes it will turn colors other than white, or have strange smells or tastes.
   e. The ordinary cause of milk souring is from bacteria with a strange, "soapy" smell, usually the color of sky blue.

11. The author of the passage would most likely agree most with which of the following?

   a. Milk researchers in the past have been incompetent and have sent us on a wild goose chase when determining what causes milk souring.

   b. Dairymen are considered more expert in the field of milk souring than milk researchers.

   c. The study of milk souring has improved throughout the years, as we now understand more of what causes milk souring and what happens afterward.

   d. Any type of bacteria will turn milk sour, so it's best to keep milk in an airtight container while it is being used.

   e. The effects of milk souring is a natural occurrence of milk, so it should not be dangerous to consume.

12. Given the author's account of the consequences of milk souring, which of the following is most closely analogous to the author's description of what happens after milk becomes slimy?

   a. The chemical change that occurs when a firework explodes.

   b. A rainstorm that overwaters a succulent plant.

   c. Mercury inside of a thermometer that leaks out.

   d. A child who swallows flea medication.

   e. A large block of ice that melts into a liquid.

13. What type of paragraph would most likely come after the third paragraph in the passage?

   a. A paragraph depicting the general effects of bacteria on milk.

   b. A paragraph explaining a broad history of what researchers have found in regard to milk souring.

   c. A paragraph outlining the properties of milk souring and the way in which it occurs.

   d. A paragraph showing the ways bacteria infiltrate milk and ways to avoid this infiltration.

   e. A paragraph naming all the bacteria in alphabetical order with a brief definition of what each does to milk.

Questions 14-20 are based on the following two passages, labeled "Passage A" and "Passage B":

Passage A

(from "Free Speech in War Time" by James Parker Hall, written in 1921, published in Columbia Law Review, Vol. 21 No. 6)

In approaching this problem of interpretation, we may first put out of consideration certain obvious limitations upon the generality of all guaranties of free speech. An occasional unthinking malcontent may urge that the only meaning not fraught with danger to liberty is the literal one that no utterance may be forbidden, no matter what its intent or result; but in fact, it is nowhere seriously argued by anyone whose opinion is entitled to respect that direct and intentional incitations to crime may not be forbidden by the state. If a state may properly forbid murder or robbery or treason, it may also punish those who induce or counsel the commission of such crimes. Any other view makes a mockery of the state's power to declare and punish offences. And what the state may do to prevent the incitement of serious crimes that are universally condemned, it may also do to prevent the incitement of lesser crimes, or of those in regard to the bad tendency of which public opinion is divided. That is, if the state may punish John for burning straw in an alley, it may also constitutionally punish Frank for inciting John to do it, though Frank did so by speech or writing. And if, in 1857, the United States could punish John for helping a fugitive slave to escape, it could also punish Frank for inducing John to do this, even though a large section of public opinion might applaud John and condemn the Fugitive Slave Law.

Passage B

(from "Freedom of Speech in War Time" by Zechariah Chafee, Jr. written in 1919, published in Harvard Law Review Vol. 32 No. 8)

The true boundary line of the First Amendment can be fixed only when Congress and the courts realize that the principle on which speech is classified as lawful or unlawful involves the balancing against each other of two very important social interests, in public safety and in the search for truth. Every reasonable attempt should be made to maintain both interests unimpaired, and the great interest in free speech should be sacrificed only when the interest in public safety is really imperiled, and not, as most men believe, when it is barely conceivable that it may be slightly affected. In war time, therefore, speech should be unrestricted by the censorship or by punishment, unless it is clearly liable to cause direct and dangerous interference with the conduct of the war.

Thus our problem of locating the boundary line of free speech is solved. It is fixed close to the point where words will give rise to unlawful acts. We cannot define the right of free speech with the precision of the Rule against Perpetuities or the Rule in Shelley's Case, because it involves national policies which are much more flexible than private property, but we can establish a workable principle of classification in this method of balancing and this broad test of certain danger. There is a similar balancing in the determination of what is "due process of law." And we can with certitude declare that the First Amendment forbids the punishment of words merely for their injurious tendencies. The history of the Amendment and the political function of free speech corroborate each other and make this conclusion plain.

14. Which one of the following questions is central to both passages?
    a. Why is freedom of speech something to be protected in the first place?
    b. Do people want absolute liberty or do they only want liberty for a certain purpose?
    c. What is the true definition of freedom of speech in a democracy?
    d. How can we find an appropriate boundary of freedom of speech during wartime?
    e. What is the interpretation of the first amendment and its limitations?

15. The authors of the two passages would be most likely to disagree over which of the following?
    a. A man is thrown in jail due to his provocation of violence in Washington D.C. during a riot.
    b. A man is thrown in jail for stealing bread for his starving family, and the judge has mercy for him and lets him go.
    c. A man is thrown in jail for encouraging a riot against the U.S. government for the wartime tactics although no violence ensues.
    d. A man is thrown in jail because he has been caught as a German spy working within the U.S. army.
    e. A man is thrown in jail because he murdered a German-born citizen whom he thought was working for the Central Powers during World War I.

16. The relationship between Passage A and Passage B is most analogous to the relationship between the documents described in which of the following?

    a. A research report that asserts water pollution in major cities in California has increased by thirty percent in the past five years; an article advocating the cessation of chicken farms in California near rivers to avoid pollution.

    b. An article detailing the effects of radiation in Fukushima; a research report describing the deaths and birth defects as a result of the hazardous waste dumped on the Somali Coast.

    c. An article that suggests that labor laws during times of war should be left up to the states; an article that showcases labor laws during the past that have been altered due to the current crisis of war.

    d. A research report arguing that the leading cause of methane emissions in the world is from agriculture practices; an article citing that the leading cause of methane emissions in the world is from the transportation of coal, oil, and natural gas.

    e. A journal article in the Netherlands about the law of euthanasia that cites evidence to support only the act of passive euthanasia as an appropriate way to die; a journal article in the Netherlands about the law of euthanasia that cites evidence to support voluntary euthanasia in any aspect.

17. The author uses the examples in the last lines of Passage A in order to do what?

    a. To demonstrate different types of crimes for the purpose of comparing them to see by which one the principle of freedom of speech would become objectionable.

    b. To demonstrate that anyone who incites a crime, despite the severity or magnitude of the crime, should be held accountable for that crime in some degree.

    c. To prove that the definition of "freedom of speech" is altered depending on what kind of crime is being committed.

    d. To show that some crimes are in the best interest of a nation and should not be punishable if they are proven to prevent harm to others.

    e. To suggest that the crimes mentioned should be reopened in order to punish those who incited the crimes.

18. Which of the following, if true, would most seriously undermine the claim proposed by the author in Passage A that if the state can punish a crime, then it can punish the incitement of that crime?

    a. The idea that human beings are able and likely to change their mind between the utterance and execution of an event that may harm others.

    b. The idea that human beings will always choose what they think is right based on their cultural upbringing.

    c. The idea that the limitation of free speech by the government during wartime will protect the country from any group that causes a threat to that country's freedom.

    d. The idea that those who support freedom of speech probably have intentions of subverting the government.

    e. The idea that if a man encourages a woman to commit a crime and she succeeds, the man is just as guilty as the woman.

19. What is the primary purpose of the second passage?
   a. To analyze the First Amendment in historical situations in order to make an analogy to the current war at hand in the nation.
   b. To demonstrate that the boundaries set during wartime are different from that when the country is at peace, and that we should change our laws accordingly.
   c. To offer the idea that during wartime, the principle of freedom of speech should be limited to that of even minor utterances in relation to a crime.
   d. To call upon the interpretation of freedom of speech to be already evident in the First Amendment and to offer a clear perimeter of the principle during war time.
   e. To assert that any limitation on freedom of speech is a violation of human rights and that the circumstances of war do not change this violation.

20. Which of the following words, if substituted for the word *malecontent* in Passage A, would LEAST change the meaning of the sentence?
   a. Regimen
   b. Cacophony
   c. Anecdote
   d. Residua
   e. Grievance

Questions 21-25 are based on the following passage from Rhetoric and Poetry in the Renaissance: A Study of Rhetorical Terms in English Renaissance Literary Criticism *by DL Clark:*

To the Greeks and Romans, rhetoric meant the theory of oratory. As a pedagogical mechanism, it endeavored to teach students to persuade an audience. The content of rhetoric included all that the ancients had learned to be of value in persuasive public speech. It taught how to work up a case by drawing valid inferences from sound evidence, how to organize this material in the most persuasive order, and how to compose in clear and harmonious sentences. Thus, to the Greeks and Romans, rhetoric was defined by its function of discovering means to persuasion and was taught in the schools as something that every free-born man could and should learn.

In both these respects the ancients felt that poetics, the theory of poetry, was different from rhetoric. As the critical theorists believed that the poets were inspired, they endeavored less to teach men to be poets than to point out the excellences which the poets had attained. Although these critics generally, with the exceptions of Aristotle and Eratosthenes, believed the greatest value of poetry to be in the teaching of morality, no one of them endeavored to define poetry, as they did rhetoric, by its purpose. To Aristotle, and centuries later to Plutarch, the distinguishing mark of poetry was imitation. Not until the renaissance did critics define poetry as an art of imitation endeavoring to inculcate morality . . .

The same essential difference between classical rhetoric and poetics appears in the content of classical poetics. Whereas classical rhetoric deals with speeches which might be delivered to convict or acquit a defendant in the law court, or to secure a certain action by the deliberative assembly, or to adorn an occasion, classical poetic deals with lyric, epic, and drama. It is a commonplace that classical literary critics paid little attention to the lyric. It is less frequently realized that they devoted almost as little space to discussion of metrics. By far the greater bulk of classical treatises on poetics is devoted to characterization and to the technique of plot construction, involving as it does narrative and dramatic unity and movement as distinct from logical unity and movement.

21. What does the author say about one way in which the purpose of poetry changed for later philosophers?

    a. The author says that at first, poetry was not defined by its purpose but was valued for its ability to be used to teach morality. Later, some philosophers would define poetry by its ability to instill morality. Finally, during the renaissance, poetry was believed to be an imitative art, but was not necessarily believed to instill morality in its readers.

    b. The author says that the classical understanding of poetry dealt with its ability to be used to teach morality. Later, philosophers would define poetry by its ability to imitate life. Finally, during the renaissance, poetry was believed to be an imitative art that instilled morality in its readers.

    c. The author says that at first, poetry was thought to be an imitation of reality, then later, philosophers valued poetry more for its ability to instill morality.

    d. The author says that the classical understanding of poetry was that it dealt with the search for truth through its content; later, the purpose of poetry would be through its entertainment value.

    e. The author says that the initial understanding of the purpose of poetry was its entertainment value. Then, as poetry evolved into a more religious era, the renaissance, it was valued for its ability to instill morality through its teaching.

22. What does the author of the passage say about classical literary critics in relation to poetics?

    a. That rhetoric was valued more than poetry because rhetoric had a definitive purpose to persuade an audience, and poetry's wavering purpose made it harder for critics to teach.

    b. That although most poetry was written as lyric, epic, or drama, the critics were most focused on the techniques of lyric and epic and their performance of musicality and structure.

    c. That although most poetry was written as lyric, epic, or drama, the critics were most focused on the techniques of the epic and drama and their performance of structure and character.

    d. That the study of poetics was more pleasurable than the study of rhetoric due to its ability to assuage its audience, and the critics, therefore, focused on what poets did to create that effect.

    e. That since poetics was made by the elite in Greek and Roman society, literary critics resented poetics for its obsession of material things and its superfluous linguistics.

23. What is the primary purpose of this passage?

    a. To alert the readers to Greek and Roman culture regarding poetic texts and the focus on characterization and plot construction rather than lyric and meter.

    b. To inform the readers of the changes in poetic critical theory throughout the years and to contrast those changes to the solidity of rhetoric.

    c. To educate the audience on rhetoric by explaining the historical implications of using rhetoric in the education system.

    d. To convince the audience that poetics is a subset of rhetoric as viewed by the Greek and Roman culture.

    e. To contemplate the differences between classical rhetoric and poetry and to consider their purposes in a particular culture.

24. The word *inculcate* in the second paragraph can be best interpreted as meaning which one of the following?

    a. Imbibe

    b. Instill

    c. Implode

    d. Inquire

    e. Idolize

25. Which of the following most closely resembles the way in which the passage is structured?
   a. The first paragraph presents an issue. The second paragraph offers a solution to the problem. The third paragraph summarizes the first two paragraphs.
   b. The first paragraph presents definitions and examples of a particular subject. The second paragraph presents a second subject in the same way. The third paragraph offers a contrast of the two subjects.
   c. The first paragraph presents an inquiry. The second paragraph explains the details of that inquiry. The last paragraph offers a solution.
   d. The first paragraph presents two subjects alongside definitions and examples. The second paragraph presents us a comparison of the two subjects. The third paragraph presents a contrast of the two subjects.
   e. The first paragraph offers a solution to a problem. The second paragraph questions the solution. The third paragraph offers a different solution.

Questions 26-29 are based on the following passage from Oregon, Washington, and Alaska. Sights and Scenes for the Tourist, *written by E.L. Lomax in 1890:*

Portland is a very beautiful city of 60,000 inhabitants and situated on the Willamette river twelve miles from its junction with the Columbia. It is perhaps true of many of the growing cities of the West, that they do not offer the same social advantages as the older cities of the East. But this is principally the case as to what may be called boom cities, where the larger part of the population is of that floating class which follows in the line of temporary growth for the purposes of speculation, and in no sense applies to those centers of trade whose prosperity is based on the solid foundation of legitimate business. As the metropolis of a vast section of country, having broad agricultural valleys filled with improved farms, surrounded by mountains rich in mineral wealth, and boundless forests of as fine timber as the world produces, the cause of Portland's growth and prosperity is the trade which it has as the center of collection and distribution of this great wealth of natural resources, and it has attracted, not the boomer and speculator, who find their profits in the wild excitement of the boom, but the merchant, manufacturer, and investor, who seek the surer if slower channels of legitimate business and investment. These have come from the East, most of them within the last few years. They came as seeking a better and wider field to engage in the same occupations they had followed in their Eastern homes, and bringing with them all the love of polite life which they had acquired there, have established here a new society, equaling in all respects that which they left behind. Here are as fine churches, as complete a system of schools, as fine residences, as great a love of music and art, as can be found at any city of the East of equal size.

But while Portland may justly claim to be the peer of any city of its size in the United States in all that pertains to social life, in the attractions of beauty of location and surroundings it stands without its peer. The work of art is but the copy of nature. What the residents of other cities see but in the copy, or must travel half the world over to see in the original, the resident of Portland has at its very door.

The city is situate on a gently-sloping ground, with, on the one side, the river, and on the other a range of hills, which, within easy walking distance, rise to an elevation of a thousand feet above the river, affording a most picturesque building site. From the very streets of the thickly settled

189

portion of the city, the Cascade Mountains, with the snow-capped peaks of Hood, Adams, St. Helens, and Rainier, are in plain view.

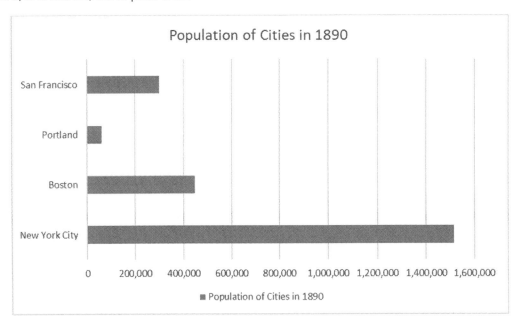

26. What is a characteristic of a "boom city," as indicated by the passage?
   a. A city that is built on solid business foundation of mineral wealth and farming.
   b. An area of land on the west coast that quickly becomes populated by residents from the east coast.
   c. A city that, due to the hot weather and dry climate, catches fire frequently, resulting in a devastating population drop.
   d. A city whose population is made up of people who seek quick fortunes rather than building a solid business foundation.
   e. A city that has been through a nuclear accident and that is inhabitable due to unlivable conditions, especially radiation.

27. By stating that "they do not offer the same social advantages as the older cities of the East" in the first paragraph, the author most likely intends to suggest that:
   a. Inhabitants who reside in older cities in the East are much more social that inhabitants who reside in newer cities in the West because of background and experience.
   b. Cities in the West have no culture compared to the East because the culture in the East comes from European influence.
   c. Cities in the East are older than cities in the West, and older cities always have better culture than newer cities.
   d. Since cities in the West are newly established, it takes them a longer time to develop cultural roots and societal functions than those cities that are already established in the East.
   e. Western cities are uninhabitable because they do not provide schools, churches, or communities wherein people learn to socialize.

28. Based on the information at the end of paragraph 1, what would the author say of Portland?
    a. It has twice as much culture as the cities in the East.
    b. It has as much culture as the cities in the East.
    c. It doesn't have as much culture as cities in the East.
    d. It doesn't have as much culture as cities in the West.
    e. It is completely void of any kind of culture.

29. How many more citizens did San Francisco have than Portland in 1890?
    a. Approximately 240,000
    b. Approximately 500,000
    c. Approximately 1,000,000
    d. Approximately 1,500,000
    e. Approximately 2,000,000

Questions 30-33 are based on the excerpt from Variation of Animals and Plants *by Charles Darwin:*

> Peach (*Amygdalus persica*).—In the last chapter I gave two cases of a peach-almond and a double-flowered almond which suddenly produced fruit closely resembling true peaches. I have also given many cases of peach-trees producing buds, which, when developed into branches, have yielded nectarines. We have seen that no less than six named and several unnamed varieties of the peach have thus produced several varieties of nectarine. I have shown that it is highly improbable that all these peach-trees, some of which are old varieties, and have been propagated by the million, are hybrids from the peach and nectarine, and that it is opposed to all analogy to attribute the occasional production of nectarines on peach-trees to the direct action of pollen from some neighbouring nectarine-tree. Several of the cases are highly remarkable, because, firstly, the fruit thus produced has sometimes been in part a nectarine and in part a peach; secondly, because nectarines thus suddenly produced have reproduced themselves by seed; and thirdly, because nectarines are produced from peach-trees from seed as well as from buds. The seed of the nectarine, on the other hand, occasionally produces peaches; and we have seen in one instance that a nectarine-tree yielded peaches by bud-variation. As the peach is certainly the oldest or primary variety, the production of peaches from nectarines, either by seeds or buds, may perhaps be considered as a case of reversion. Certain trees have also been described as indifferently bearing peaches or nectarines, and this may be considered as bud-variation carried to an extreme degree.
>
> The grosse mignonne peach at Montreuil produced "from a sporting branch" the grosse mignonne tardive, "a most excellent variety," which ripens its fruit a fortnight later than the parent tree, and is equally good. This same peach has likewise produced by bud-variation the early grosse mignonne. Hunt's large tawny nectarine "originated from Hunt's small tawny nectarine, but not through seminal reproduction."

30. Which statement is NOT a detail from the passage?
    a. At least six named varieties of the peach have produced several varieties of nectarine.
    b. It is not probable that all of the peach-trees mentioned are hybrids from the peach and nectarine.
    c. An unremarkable case is the fact that nectarines are produced from peach-trees from seed as well as from buds.
    d. The production of peaches from nectarines might be considered a case of reversion.
    e. Some trees have been described as indifferently bearing peaches or nectarines.

31. What is the meaning of the word "propagated" in the first paragraph of this passage?
    a. Multiplied
    b. Diminished
    c. Watered
    d. Uprooted
    e. Eradicated

32. Which of the following most closely reveals the author's tone in this passage?
    a. Enthusiastic
    b. Objective
    c. Critical
    d. Desperate
    e. Persuasive

33. Which of the following is an accurate paraphrasing of the following phrase?

    Certain trees have also been described as indifferently bearing peaches or nectarines, and this may be considered as bud-variation carried to an extreme degree.

    a. Some trees are described as bearing peaches and some trees have been described as bearing nectarines, but individually the buds are extreme examples of variation.
    b. One way in which bud-variation is said to be carried to an extreme degree is when specific trees have been shown to casually produce peaches or nectarines.
    c. Certain trees are indifferent to bud-variation, as recently shown in the trees that produce both peaches and nectarines in the same season.
    d. Nectarines and peaches are known to have cross-variation in their buds, which indifferently bears other sorts of fruit to an extreme degree.
    e. The bud variation of apples and nectarines is shown in certain trees to an extreme degree once they have indifferently produced the fruit.

Questions 34-37 are based on the excerpt from A Christmas Carol *by Charles Dickens:*

    Meanwhile the fog and darkness thickened so, that people ran about with flaring links, proffering their services to go before horses in carriages, and conduct them on their way. The ancient tower of a church, whose gruff old bell was always peeping slily down at Scrooge out of a Gothic window in the wall, became invisible, and struck the hours and quarters in the clouds, with tremulous vibrations afterwards as if its teeth were chattering in its frozen head up there. The cold became intense. In the main street, at the corner of the court, some labourers were repairing the gas-pipes, and had lighted a great fire in a brazier, round which a party of ragged men and boys were gathered: warming their hands and winking their eyes before the blaze in rapture. The water-plug being left in solitude, its overflowings sullenly congealed, and turned to misanthropic ice. The brightness of the shops where holly sprigs and berries crackled in the lamp heat of the windows, made pale faces ruddy as they passed. Poulterers' and grocers' trades became a splendid joke; a glorious pageant, with which it was next to impossible to believe that such dull principles as bargain and sale had anything to do. The Lord Mayor, in the stronghold of the mighty Mansion House, gave orders to his fifty cooks and butlers to keep Christmas as a Lord Mayor's household should; and even the little tailor, whom he had fined five shillings on the previous Monday for being drunk and bloodthirsty in the streets, stirred up to-morrow's pudding in his garret, while his lean wife and the baby sallied out to buy the beef.

Foggier yet, and colder. Piercing, searching, biting cold. If the good Saint Dunstan had but nipped the Evil Spirit's nose with a touch of such weather as that, instead of using his familiar weapons, then indeed he would have roared to lusty purpose. The owner of one scant young nose, gnawed and mumbled by the hungry cold as bones are gnawed by dogs, stopped down at Scrooge's keyhole to regale him with a Christmas carol: but at the first sound of

"God bless you, merry gentleman! May nothing you dismay!"

Scrooge seized the ruler with such energy of action, that the singer fled in terror, leaving the keyhole to the fog and even more congenial frost.

34. In the context in which it appears, *congealed* most nearly means which of the following?
    a. Burst
    b. Loosened
    c. Shrank
    d. Thickened
    e. Carried

35. Which of the following can NOT be inferred from the passage?
    a. The season of this narrative is in the winter time.
    b. The majority of the narrative is located in a bustling city street.
    c. This passage takes place during the night time.
    d. The Lord Mayor is a wealthy person within the narrative.
    e. Scrooge is not in the best mood in the passage.

36. According to the passage, the poulterers and grocers
    a. were so poor in the quality of their products that customers saw them as a joke.
    b. put on a pageant in the streets every year for Christmas to entice their customers.
    c. did not believe in Christmas, so they refused to participate in the town parade.
    d. set their shops up to be entertaining public spectacles rather than a dull trade exchange.
    e. were somber in their selling tactics and treated customers with respect.

37. The author's depiction of the scene in the last few paragraphs does all of the following EXCEPT:
    a. Offer an allusion to religious affiliation in England.
    b. Attempt to evoke empathy for the character of Scrooge.
    c. Provide a palpable experience through the use of imagery and diction.
    d. Depict Scrooge as an uncaring, terrifying character to his fellows.
    e. Presents personification as a literary device.

Questions 38-40 are based on the book On the Trail *by Lina Beard and Adelia Belle Beard:*

For any journey, by rail or by boat, one has a general idea of the direction to be taken, the character of the land or water to be crossed, and of what one will find at the end. So it should be in striking the trail. Learn all you can about the path you are to follow. Whether it is plain or obscure, wet or dry; where it leads; and its length, measured more by time than by actual miles. A smooth, even trail of five miles will not consume the time and strength that must be expended upon a trail of half that length which leads over uneven ground, varied by bogs and obstructed by rocks and fallen trees, or a trail that is all up-hill climbing. If you are a novice and accustomed to walking only over smooth and level ground, you must allow more time for covering the distance than an experienced person would require and must count upon the expenditure of more strength, because your feet are not trained to the wilderness paths with their pitfalls and traps for the unwary, and every nerve and muscle will be strained to secure a safe foothold amid the tangled roots, on the slippery, moss-covered logs, over precipitous rocks that lie in your path. It will take time to pick your way over boggy places where the water oozes up through the thin, loamy soil as through a sponge; and experience alone will teach you which hummock of grass or moss will make a safe stepping-place and will not sink beneath your weight and soak your feet with hidden water. Do not scorn to learn all you can about the trail you are to take . . . It is not that you hesitate to encounter difficulties, but that you may prepare for them. In unknown regions take a responsible guide with you, unless the trail is short, easily followed, and a frequented one. Do not go alone through lonely places; and, being on the trail, keep it and try no explorations of your own, at least not until you are quite familiar with the country and the ways of the wild.

Blazing the Trail

A woodsman usually blazes his trail by chipping with his axe the trees he passes, leaving white scars on their trunks, and to follow such a trail you stand at your first tree until you see the blaze on the next, then go that and look for the one farther on; going in this way from tree to tree you keep the trail though it may, underfoot, be overgrown and indistinguishable.

If you must make a trail of your own, blaze it as you go by bending down and breaking branches of trees, underbrush, and bushes. Let the broken branches be on the side of bush or tree in the direction you are going, but bent down away from that side, or toward the bush, so that the lighter underside of the leaves will show and make a plain trail. Make these signs conspicuous and close together, for in returning, a dozen feet without the broken branch will sometimes confuse you, especially as everything has a different look when seen from the opposite side. By this same token it is a wise precaution to look back frequently as you go and impress the homeward-bound landmarks on your memory. If in your wanderings you have branched off and made ineffectual or blind trails which lead nowhere, and, in returning to camp, you are led astray by one of them, do not leave the false trail and strike out to make a new one, but turn back and follow the false trail to its beginning, for it must lead to the true trail again. Don't lose sight of your broken branches.

38. In the image, what part of the text is the girl most likely emulating?
    a. Building a trap
    b. Setting up camp
    c. Blazing the trail
    d. Picking berries to eat
    e. Fishing the stream

39. According to the passage, what does the author say about unknown regions?
    a. You should try and explore unknown regions in order to learn the land better.
    b. Unless the trail is short or frequented, you should take a responsible guide with you.
    c. All unknown regions will contain pitfalls, traps, and boggy places.
    d. It's better to travel unknown regions by rail rather than by foot.
    e. You should never take unknown regions, even with an experienced guide.

40. Which statement is NOT a detail from the passage?
    a. Learning about the trail beforehand is imperative.
    b. Time will differ depending on the land.
    c. Once you are familiar with the outdoors you can go places on your own.
    d. Be careful for wild animals on the trail you are on.
    e. You can leave broken branches to mark your path.

Answer Explanations

1. D: Neutrality due to the style of the report. The report is mostly objective; we see very little language that entails any strong emotion whatsoever. The story is told almost as an objective documentation of a sequence of actions—we see the president sitting in his box with his wife, their enjoyment of the show, Booth's walk through the crowd to the box, and Ford's consideration of Booth's movements. There is perhaps a small amount of bias when the author mentions the president's "worthy wife." However, the word choice and style show no signs of excitement, sadness, anger, or apprehension from the author's perspective, so the best answer is Choice *D*.

2. B: Mr. Ford assumed Booth's movement throughout the theater was due to being familiar with the theater. Choice *A* is incorrect; although Booth does eventually make his way to Lincoln's box, Mr. Ford does not make this distinction in this part of the passage. Choice *C* is incorrect; although the passage mentions "companions," it mentions Lincoln's companions rather than Booth's companions. Choice *D* is incorrect; the passage mentions "dress circle," which means the first level of the theater, but this is different from a "dressing room." Finally, Choice *E* is incorrect; the passage mentions a "symptom" but does not signify a symptom from an illness.

3. C: A lead singer leaves their band to begin a solo career, and the band's sales on their next album drop by 50 percent. The original source of the analogy displays someone significant to an event who leaves, and then the event becomes the worst for it. We see Mr. Sothern leaving the theater company, and then the play becoming a "very dull affair." Choice *A* depicts a dancer who backs out of an event before the final performance, so this is incorrect. Choice *B* shows a basketball player leaving an event, and then the team makes it to the championship but then loses. This choice could be a contestant for the right answer; however, we don't know if the team has become the worst for his departure or the better for it. We simply do not have enough information here. Choice *D* is incorrect. The actor departs an event, but there is no assessment of the quality of the movie. It simply states what actors filled in instead. Choice *E* is incorrect because the opposite of the source happens; the professor leaves the entity, and the entity becomes better. Additionally, the betterment of the entity is not due to the individual leaving. Choice *E* is not analogous to the source.

4. A: A chronological account in a fiction novel of a woman and a man meeting for the first time. It's tempting to mark Choice A wrong because the genres are different. Choice *A* is a fiction text, and the original passage is not a fictional account. However, the question stem asks specifically for organizational structure. Choice *A* is a chronological structure just like the passage, so this is the correct answer. The passage does not have a cause and effect, problem/solution, or compare/contrast structure, making Choices *B*, *D*, and *E* incorrect. Choice *C* is tempting because it mentions an autobiography; however, the structure of this text starts at the end and works its way toward the beginning, which is the opposite structure of the original passage.

5. C: The word *adornments* would LEAST change the meaning of the sentence because it's the most closely related word to *festoons*. The other choices don't make sense in the context of the sentence. *Feathers* of flags, *armies* of flags, *buckets* of flags, and *boats* of flags are not as accurate as the word *adornments* of flags. The passage also talks about other décor in the setting, so the word *adornments* fits right in with the context of the paragraph.

6. D: The primary purpose of the passage is to recount in detail the events that led up to Abraham Lincoln's death. Choice *A* is incorrect; the author makes no claims and uses no rhetoric of persuasion towards the audience. Choice *B* is incorrect, though it's a tempting choice; the passage depicts the setting in exorbitant detail, but the setting itself is not the primary purpose of the passage. Choice *C* is incorrect; one could argue this is a narrative, and the passage is about Lincoln's last few hours, but this isn't the *best* choice. The best choice recounts the details that leads up to Lincoln's death. Finally, Choice *E* is incorrect. The author does not try to prove or disprove anything to the audience, and the passage does not even make it to when Lincoln gets shot, so this part of the story is irrelevant.

7. D: The word *deleterious* can be best interpreted as referring to the word *ruinous*. The first paragraph attempts to explain the process of milk souring, so the "acid" would probably prove "ruinous" to the growth of bacteria and cause souring. Choice *A, amicable,* means friendly, so this does not make sense in context. Choice *B, smoldering,* means to boil or simmer, so this is also incorrect. Choices *C* and *E, luminous* and *virtuous,* have positive connotations and don't make sense in the context of the passage. Luminous means shining or brilliant, and virtuous means to be honest or ethical.

8. B: The author begins by explaining a process or phenomenon, then gives the history of the study of this phenomenon, and ends by presenting the effects of this phenomenon. The author explains the process of souring in the first paragraph by informing the reader that "it is due to the action of certain of the milk bacteria upon the milk sugar which converts it into lactic acid, and this acid gives the sour taste and curdles the milk." In the second paragraph, we see how the phenomenon of milk souring was viewed when it was "first studied," and then we proceed to gain insight into "recent investigations" toward the end of the paragraph. Finally, the passage ends by presenting the effects of the phenomenon of milk souring. We see the milk curdling, becoming bitter, tasting soapy, turning blue, or becoming thread-like. All of the other answer choices are incorrect.

9: A: The primary purpose is to inform the reader of the phenomenon, investigation, and consequences of milk souring. Choice *B* is incorrect because the passage states that *Bacillus acidi lactici* is not the only cause of milk souring. Choice *C* is incorrect because, although the author mentions the findings of researchers, the main purpose of the text does not seek to describe their accounts and findings, as we are not even told the names of any of the researchers. Choice *D* is tricky. We do see the author present us with new findings in contrast to the first cases studied by researchers. However, this information is only in the second paragraph, so it is not the primary purpose of the *entire passage*. Finally, Choice *E* is incorrect because the genre of the passage is more informative than narrative, although the author does talk about the phenomenon of milk souring and its subsequent effects.

10. C: Milk souring is caused mostly by a species of bacteria identical to that of *Bacillus acidi lactici* although there are a variety of other bacteria that cause milk souring as well. Choice *A* is incorrect because it contradicts the assertion that the souring is still caused by a variety of bacteria. Choice *B* is incorrect because the ordinary cause of milk souring *is known* to current researchers. Choice *D* is incorrect because this names mostly the effects of milk souring, not the cause. Choice *E* is incorrect because the bacteria itself doesn't have a strange soapy smell or is a different color, but it eventually will cause the milk to produce these effects.

11. C: The study of milk souring has improved throughout the years, as we now understand more of what causes milk souring and what happens afterward. None of the choices here are explicitly stated, so we have to rely on our ability to make inferences. Choice *A* is incorrect because there is no indication from the author that milk researchers in the past have been incompetent—only that recent research has done a better job of studying the phenomenon of milk souring. Choice *B* is incorrect because the author refers to dairymen in relation to the effects of milk souring and their "troubles" surrounding milk souring, and does not compare them to milk researchers. Choice *D* is incorrect because we are told in the second paragraph that only certain types of bacteria are able to sour milk. Choice *E* is incorrect; although we are told that milk souring is a natural occurrence, the author makes no implication that soured milk is safe to consume. Choice *C* is the best answer choice here because although the author does not directly state that the study of milk souring has improved, we can see this might be true due to the comparison of old studies to newer studies, and the fact that the newer studies are being used as a reference in the passage.

12. A: It is most analogous to the chemical change that occurs when a firework explodes. The author tells us that after milk becomes slimy, "it persists in spite of all attempts made to remedy it," which means the milk has gone through a chemical change. It has changed its state from milk to sour milk by changing its odor, color, and material. After a firework explodes, there is nothing one can do to change the substance of a firework back to its original form—the original substance is turned into sound and light. Choice *B* is incorrect because, although the rain overwatered the plant, it's possible that the plant is able to recover from this. Choice *C* is incorrect because although mercury leaking out may be dangerous, the actual substance itself stays the same and does not alter into something else. Choice *D* is incorrect; this situation is not analogous to the alteration of a substance. Choice *E* is also incorrect. Ice melting into a liquid is a physical change, which means it can be undone. Milk turning sour, as the author asserts, cannot be undone.

13. D: It would most likely be a paragraph showing the ways bacteria infiltrate milk and ways to avoid this infiltration. Choices *A, B,* and *C* are incorrect because these are already represented in the third, second, and first paragraphs. Choice *E* is incorrect; this choice isn't impossible. There could be a glossary right after the third paragraph, but this would be an awkward place for a glossary. Choice *D* is the best answer because it follows a sort of problem/solution structure in writing.

14. E: A central question to both passages is: What is the interpretation of the first amendment and its limitations? Choice *A* is incorrect; this is a question for the first passage but it does not apply to the second. Choice *B* is incorrect; a quote mentions this at the end of the first passage, but this question is not found in the second passage. Choice *C* is incorrect, as the passages are not concerned with the definition of freedom of speech, but how to interpret it. Choice *D* is incorrect; this is a question for the second passage, but it is not found in the first passage.

15. C: The authors would most likely disagree over the man thrown in jail for encouraging a riot against the U.S. government for the wartime tactics although no violence ensued. The author of Passage A says that "If a state may properly forbid murder or robbery or treason, it may also punish those who induce or counsel the commission of such crimes." This statement tells us that the author of Passage A would support throwing the man in jail for encouraging a riot, although no violence ensues. The author of Passage B states that "And we can with certitude declare that the First Amendment forbids the punishment of words merely for their injurious tendencies." This is the best answer choice because we are clear on each author's stance in this situation. Choice *A* is tricky; the author of Passage A would definitely agree with this, but it's questionable whether the author of Passage B would also agree. Violence does ensue at the capitol as a result of this man's provocation, and the author of Passage B states "speech should be unrestricted by censorship . . . unless it is clearly liable to cause direct . . . interference with the conduct of war." This answer is close, but it is not the *best* choice. Choice *B* is incorrect because we have no way of knowing what the authors' philosophies are in this situation. Choice *D* is incorrect because, again, we have no way of knowing what the authors would do in this situation, although it's assumed they would probably both agree with this. Choice *E* is something the authors would probably both agree on, because brutal violence ensued, but it has nothing to do with free speech, so we have no way of knowing for sure.

16. E: Choice *E* is the best answer. To figure out the correct answer choice we must find out the relationship between Passage A and Passage B. Between the two passages, we have a general principle (freedom of speech) that is questioned on the basis of interpretation. In Choice *E,* we see that we have a general principle (right to die, or euthanasia) that is questioned on the basis of interpretation as well. Should euthanasia only include passive euthanasia, or euthanasia in any aspect? Choice *A* is a problem/solution relationship; the first option outlines a problem, and the second option delivers a solution, so this choice is incorrect. Choice *B* is incorrect because it does not question the interpretation of a principle, but rather describes the effects of two events that happened in the past involving contamination of radioactive substances. Choice *C* begins with a principle—that of labor laws during wartime—but in the second option, the interpretation isn't questioned. The second option looks at the historical precedent of labor laws in the past during wartime. Choice *D* is incorrect because the two texts disagree over the cause of something rather than the interpretation of it.

17. B: Choice *B* is the best answer choice because the author is trying to demonstrate via the examples that anyone who incites a crime, despite the severity or magnitude of the crime, should be held accountable for that crime in some degree. Choice *A* is incorrect because the crimes mentioned are not being compared to each other, but they are being used to demonstrate a point. Choice *C* is incorrect because the author makes the same point using both of the examples and does not question the definition of freedom of speech but its ability to be limited. Choice *D* is incorrect because this sentiment goes against what the author has been arguing throughout the passage. Choice *E* is incorrect because the author does not suggest that the crimes mentioned be reopened anywhere in the passage.

18. A: The idea that human beings are able and likely to change their mind between the utterance and execution of an event that may harm others most seriously undermines the claim because it brings into question the bad tendency of a crime and points out the difference between utterance and action in moral situations. Choice *B* is incorrect; this idea does not undermine the claim at hand, but introduces an observation irrelevant to the claim. Choices *C, D,* and *E* would most likely strengthen the argument's claim; or, they are at least supported by the author in Passage A.

19. D: The primary purpose is to call upon the interpretation of freedom of speech to be already evident in the First Amendment and to offer a clear perimeter of the principle during war time. Choice *A* is incorrect; the passage calls upon no historical situations as precedent in this passage. Choice *B* is incorrect; we can infer that the author would not agree with this, because the author states that "In war time, therefore, speech should be unrestricted . . . by punishment." Choice *C* is incorrect; this is more consistent with the main idea of the first passage. Choice *E* is incorrect; the passage states a limitation in saying that "speech should be unrestricted . . . unless it is clearly liable to cause direct and dangerous interference with the conduct of war."

20. E: The word that would least change the meaning of the sentence is *grievance. Malcontent* is a complaint or grievance, and in this context would be uttered in advocation of absolute freedom of speech. Choice *A, regimen,* means a pattern of living, and would not make sense in this context. Choice *B, cacophony,* means a harsh noise or mix of discordant noises; someone may express or "urge" a cacophony but it would be an awkward word in this context. Choice *C, anecdote,* is a short account of an amusing story. Since the word is a noun, it fits grammatically inside the sentence, but anecdotes are usually thought out, and this word is considered "unthinking." Choice *D, residua,* means an outcome, and also does not make sense within this context.

21. B: The author says that the classical understanding of poetry dealt with its ability to be used to teach morality. Later, philosophers would define poetry by its ability to imitate life. Finally, during the renaissance, poetry was believed to be an imitative art that instilled morality in its readers. The rest of the answer choices improperly interpret this explanation in the passage. Poetry was never mentioned for use in entertainment, which makes Choices *D* and *E* incorrect. Choices *A* and *C* are incorrect because they mix up the chronological order.

22. C: The author says that although most poetry was written as lyric, epic, or drama, the critics were most focused on the techniques of the epic and drama and their performance of structure and character. This is the best answer choice as portrayed by paragraph three. Choice *A* is incorrect because nowhere in the passage does it say rhetoric was more valued than poetry, although it did seem to have a more definitive purpose than poetry. Choice *B* is incorrect; this almost mirrors Choice *A*, but the critics were *not* focused on the lyric, as the passage indicates. Choice *D* is incorrect because the passage does not mention that the study of poetics was more pleasurable than the study of rhetoric. Choice *E* is incorrect because again, we do not see anywhere in the passage where poetry was reserved for the most elite in society.

23. E: The purpose is to contemplate the differences between classical rhetoric and poetry and to consider their purposes in a particular culture. Choice *A* is incorrect; this thought is discussed in the third paragraph, but it is not the main idea of the passage. Choice *B* is incorrect; although changes in poetics throughout the years is mentioned, this is not the main idea of the passage. Choice *C* is incorrect; although this is partly true—that rhetoric within the education system is mentioned—the subject of poetics is left out of this answer choice. Choice *D* is incorrect; the passage makes no mention of poetics being a subset of rhetoric.

24. B: The correct answer choice is Choice *B*, *instill*. Choice *A*, *imbibe*, means to drink heavily, so this choice is incorrect. Choice *C*, *implode*, means to collapse inward, which does not make sense in this context. Choice *D*, *inquire*, means to investigate. This option is better than the other options, but it is not as accurate as *instill*. Choice *E*, *idolize*, means to admire, which does not make sense in this context.

25. B: The first paragraph presents definitions and examples of a particular subject. The second paragraph presents a second subject in the same way. The third paragraph offers a contrast of the two subjects. In the passage, we see the first paragraph defining rhetoric and offering examples of how the Greeks and Romans taught this subject. In the second paragraph, poetics is defined and examples of its dynamic definition are provided. In the third paragraph, the contrast between rhetoric and poetry is characterized through how each of these were studied in a classical context.

26. D: A city whose population is made up of people who seek quick fortunes rather than building a solid business foundation. Choice *A* is a characteristic of Portland, but not that of a boom city. Choice *B* is close—a boom city is one that becomes quickly populated, but it is not necessarily always populated by residents from the east coast. Choice *C* is incorrect because a boom city is not one that catches fire frequently, but one made up of people who are looking to make quick fortunes from the resources provided on the land. Choice *E* is incorrect because this describes a nuclear accident, such as what happened at Chernobyl, but this type of accident it not deemed a "boom city."

27. D: Choice *D* is the best answer because of the surrounding context. We can see that the fact that Portland is a "boom city" means that the "floating class" go through, a group of people who only have temporary roots put down. This would cause the main focus of the city to be on employment and industry, rather than society and culture. Choice *A* is incorrect, as we are not told about the inhabitants being social or antisocial. Choice *B* is incorrect because the text does not talk about the culture in the East regarding European influence. Choice *C* is incorrect; this is an assumption that has no evidence in the text to back it up. Choice *E* uses extreme language; saying that "Western cities are uninhabitable" is incorrect and is not an opinion of the author.

28. B: The author would say that it has as much culture as the cities in the East. The author says that Portland has "as fine churches, as complete a system of schools, as fine residences, as great a love of music and art, as can be found at any city of the East of equal size," which proves that the culture is similar in this particular city to the cities in the East.

29. A: Approximately 240,000. We know from the image that San Francisco has around 300,000 inhabitants at this time. From the text (and from the graph) we can see that Portland has 60,000 inhabitants. Subtract these two numbers to come up with 240,000.

30. C: Choice *C* is the correct answer because the word "unremarkable" should be changed to "remarkable" in order for it to be consistent with the details of the passage. This question requires close attention to the passage. Choice *A* can be found where the passage says "no less than six named and several unnamed varieties of the peach have thus produced several varieties of nectarine, so this choice is incorrect. Choice *B* can be found where the passage says "it is highly improbable that all these peach-trees . . . are hybrids from the peach and nectarine." Choice *D* is incorrect because we see in the passage that "the production of peaches from nectarines, either by seeds or buds, may perhaps be considered as a case of reversion." Choice *E* is incorrect; this detail is represented in the last sentence.

31. A: The word *multiplied* is synonymous with the word *propagated*, making Choice *A* correct. Choice *B* is incorrect because *diminished* means to decrease or recede and is the opposite of *propagated*. Choice *C* is also incorrect; *watered* is closer, because it pertains to the growth of trees, but it is not exactly the same thing as *propagated*. Choice *D* is incorrect; *uprooted* could also pertain to trees, but this answer is incorrect. Choice *E* is incorrect; *eradicated* means to be destroyed, which is opposite of *propagate*.

32. B: The author's tone in this passage can be considered objective. An objective tone means that the author is open-minded and detached about the subject. Most scientific articles are objective. Choices *A, C, D,* and *E* are incorrect. The author is not very enthusiastic on the paper; the author is not critical, but rather interested in the topic. The author is not desperate in any way here. The author is not persuasive to the audience and presents their evidence unbiasedly.

33. B: Choice *B* is the correct answer because the meaning holds true even if the words have been switched out or rearranged some. Choice *A* is incorrect because it has trees either bearing peaches or nectarines, and the trees in the original phrase bear both. Choice *C* is incorrect because the statement does not say these trees are "indifferent to bud-variation," but that they have "indifferently [bore] peaches or nectarines." Choice *D* is incorrect; the statement may use some of the same words, but the meaning is skewed in this sentence. Choice *E* is incorrect because the sentence uses the examples of "apple" rather than "peach."

34. D: *Congealed* in this context most nearly means *thickened*, because we see liquid turning into ice. Choice *B, loosened,* is the opposite of the correct answer. Choices *A, C,* and *E, burst, shrank,* and *carried,* are also incorrect.

35. C: We can infer that the season of this narrative is in the winter time. Some of the evidence here is that "the cold became intense," and people were decorating their shops with "holly sprigs,"—a Christmas tradition. It also mentions that it's Christmastime at the end of the passage. Choice *A* is incorrect. We cannot infer that the passage takes place during the night time. While we do have a statement that says that the darkness thickened, this is the only evidence we have. The darkness could be thickening because it is foggy outside. We don't have enough proof to infer this otherwise. Choice *B* is incorrect; we *can* infer that the narrative is located in a bustling city street by the actions in the story. People are running around trying to sell things, the atmosphere is busy, there is a church tolling the hours, etc. The scene switches to the Mayor's house at the end of the passage, but the answer says "majority," so this is still incorrect. Choice *D* is incorrect; we *can* infer that the Lord Mayor is wealthy—he lives in the "Mansion House" and has fifty cooks. Choice *E* is incorrect because we *can* infer that Scrooge was not kind nor receptive since he scared the caroler away.

36. D: The passage tells us that the poulterers' and grocers' trades were "a glorious pageant, with which it was next to impossible to believe that such dull principles as bargain and sale had anything to do," which means they set up their shops to be entertaining public spectacles in order to increase sales. Choice *A* is incorrect; although the word "joke" is used, it is meant to be used as a source of amusement rather than something made in poor quality. Choice *B* is incorrect; that they put on a "pageant" is figurative for the public spectacle they made with their shops, not a literal play. Choice *C* is incorrect, as this is not mentioned anywhere in the passage. Choice *E* is incorrect; *somber* means *serious*, and we are told that the trades were not *dull*, but *entertaining*, so the word *somber* incorrectly describes the trading.

37. B: The author, at least in the last few paragraphs, does not attempt to evoke empathy for the character of Scrooge. We see Scrooge lashing out at an innocent, cold boy, with no sign of affection or feeling for his harsh conditions. We see Choice *A* when the author talks about Saint Dunstan. We see Choice *C* providing a palpable experience, especially with the "piercing, searching, biting cold," among other statements. We see Choice *D* when Scrooge chases the young boy away. Choice *E* is also incorrect because the paragraph *does* present personification: "gnawed and mumbled by the hungry cold" endows the noun "cold" with humanistic features in this passage.

38. C: The section that is talked about in the text is blazing the trail, Choice *C*. The passage states that one must blaze the trail by "bending down and breaking branches of trees, underbrush, and bushes." The girl in the image is bending a branch in order to break it so that she won't get lost by "blazing the trail."

39. B: Choice *B* is the best answer here; the sentence states, "In unknown regions take a responsible guide with you, unless the trail is short, easily followed, and a frequented one." Choice *A* is incorrect; the passage does not state that you should try and explore unknown regions. Choice *C* is incorrect; the passage talks about trails that contain pitfalls, traps, and boggy places, but it does not say that *all* unknown regions contain these things. Choice *D* is incorrect; the passage mentions "rail" and "boat" as means of transport at the beginning, but it does not suggest it is better to travel unknown regions by rail. Choice *E* is incorrect; the passage suggests taking a responsible guide into unknown regions.

40. D: Choice *D* is correct; it may be real advice an experienced hiker would give to an inexperienced hiker. However, the question asks about details in the passage, and this is not in the passage. Choice *A* is incorrect; we do see the author encouraging the reader to learn about the trail beforehand . . . "wet or dry; where it leads; and its length." Choice *B* is also incorrect, because we do see the author telling us the time will lengthen with boggy or rugged places opposed to smooth places. Choice *C* is incorrect; at the end of the passage, the author tells us "do not go alone through lonely places . . . unless you are quite familiar with the country and the ways of the wild." Choice *E* is also incorrect, because the detail of this information is in the passage: "If you must make a trail of your own, blaze it as you go by bending down and breaking branches of trees, underbrush, and bushes."

Verbal

Synonyms

Word Knowledge

The Synonym section on the Upper Level SSAT is designed to assess the test taker's vocabulary knowledge and skill in determining the answer choice with a meaning that most nearly matches that of the word presented in the question. In this way, these questions task test takers with applying their vocabulary skills, understanding of language, and logic to select the best synonyms.

Question Format

The questions in this section are constructed very simply: the prompt is a single word, which is then followed by the five answer choices, each of which is also a single word. Test takers must consider the definition or meaning of the word provided in the prompt, and then select the answer choice that most nearly means the same thing. Consider the following example of a question:

SERENE
   a. Serious
   b. Calm
   c. Tired
   d. Nervous
   e. Jealous

Test takers must consider the meaning of the given word (*serene),* read the five potential definitions or possible synonyms, and pick the word that most closely means the same thing as the word in the prompt. In this case, Choice *B* is the best answer because *serene* means calm.

Analyzing Word Parts

By learning some of the etymologies of words and their parts, readers can break new words down into components and analyze their combined meanings. For example, the root word *soph* is Greek for wise or knowledge. Knowing this informs the meanings of English words including *sophomore, sophisticated,* and *philosophy.* Those who also know that *phil* is Greek for love will realize that *philosophy* means the love of knowledge. They can then extend this knowledge of *phil* to understand *philanthropist* (one who loves people), *bibliophile* (book lover), *philharmonic* (loving harmony), *hydrophilic* (water-loving), and so on. In addition, *phob-* derives from the Greek *phobos,* meaning fear. This informs all words ending with it as meaning fear of various things: *acrophobia* (fear of heights), *arachnophobia* (fear of spiders), *claustrophobia* (fear of enclosed spaces), *ergophobia* (fear of work), and *hydrophobia* (fear of water), among others.

Some words that originate from other languages, like ancient Greek, are found in large numbers and varieties of English words. An advantage of the shared ancestry of these words is that once readers recognize the meanings of some Greek words or word roots, they can determine or at least get an idea of what many different English words mean. As an example, the Greek word *métron* means to measure, a measure, or something used to measure; the English word meter derives from it. Knowing this informs many other English words, including *altimeter, barometer, diameter, hexameter, isometric,* and *metric.*

While readers must know the meanings of the other parts of these words to decipher their meaning fully, they already have an idea that they are all related in some way to measures or measuring.

While all English words ultimately derive from a proto-language known as Indo-European, many of them historically came into the developing English vocabulary later, from sources like the ancient Greeks' language, the Latin used throughout Europe and much of the Middle East during the reign of the Roman Empire, and the Anglo-Saxon languages used by England's early tribes. In addition to classic revivals and native foundations, by the Renaissance era, other influences included French, German, Italian, and Spanish. Today we can often discern English word meanings by knowing common roots and affixes, particularly from Greek and Latin.

The following is a list of common prefixes and their meanings:

Prefix	Definition	Examples
a-	without	atheist, agnostic
ad-	to, toward	advance
ante-	before	antecedent, antedate
anti-	opposing	antipathy, antidote
auto-	self	autonomy, autobiography
bene-	well, good	benefit, benefactor
bi-	two	bisect, biennial
bio-	life	biology, biosphere
chron-	time	chronometer, synchronize
circum-	around	circumspect, circumference
com-	with, together	commotion, complicate
contra-	against, opposing	contradict, contravene
cred-	belief, trust	credible, credit
de-	from	depart
dem-	people	demographics, democracy
dis-	away, off, down, not	dissent, disappear
equi-	equal, equally	equivalent
ex-	former, out of	extract
for-	away, off, from	forget, forswear
fore-	before, previous	foretell, forefathers
homo-	same, equal	homogenized
hyper-	excessive, over	hypercritical, hypertension
in-	in, into	intrude, invade
inter-	among, between	intercede, interrupt
mal-	bad, poorly, not	malfunction
micr-	small	microbe, microscope
mis-	bad, poorly, not	misspell, misfire
mono-	one, single	monogamy, monologue
mor-	die, death	mortality, mortuary
neo-	new	neolithic, neoconservative
non-	not	nonentity, nonsense
omni-	all, everywhere	omniscient

over-	above	overbearing
pan-	all, entire	panorama, pandemonium
para-	beside, beyond	parallel, paradox
phil-	love, affection	philosophy, philanthropic
poly-	many	polymorphous, polygamous
pre-	before, previous	prevent, preclude
prim-	first, early	primitive, primary
pro-	forward, in place of	propel, pronoun
re-	back, backward, again	revoke, recur
sub-	under, beneath	subjugate, substitute
super-	above, extra	supersede, supernumerary
trans-	across, beyond, over	transact, transport
ultra-	beyond, excessively	ultramodern, ultrasonic, ultraviolet
un-	not, reverse of	unhappy, unlock
vis-	to see	visage, visible

The following is a list of common suffixes and their meanings:

Suffix	Definition	Examples
-able	likely, able to	capable, tolerable
-ance	act, condition	acceptance, vigilance
-ard	one that does excessively	drunkard, wizard
-ation	action, state	occupation, starvation
-cy	state, condition	accuracy, captaincy
-er	one who does	teacher
-esce	become, grow, continue	convalesce, acquiesce
-esque	in the style of, like	picturesque, grotesque
-ess	feminine	waitress, lioness
-ful	full of, marked by	thankful, zestful
-ible	able, fit	edible, possible, divisible
-ion	action, result, state	union, fusion
-ish	suggesting, like	churlish, childish
-ism	act, manner, doctrine	barbarism, socialism
-ist	doer, believer	monopolist, socialist
-ition	action, result, state,	sedition, expedition
-ity	quality, condition	acidity, civility
-ize	cause to be, treat with	sterilize, mechanize, criticize
-less	lacking, without	hopeless, countless
-like	like, similar	childlike, dreamlike
-ly	like, of the nature of	friendly, positively
-ment	means, result, action	refreshment, disappointment
-ness	quality, state	greatness, tallness
-or	doer, office, action	juror, elevator, honor
-ous	marked by, given to	religious, riotous
-some	apt to, showing	tiresome, lonesome

-th	act, state, quality	warmth, width
-ty	quality, state	enmity, activity

The following is a list of root words and their meanings:

Root	Definition	Examples
ambi	both	ambidextrous, ambiguous
anthropo	man; humanity	anthropomorphism, anthropology
auto	self	automobile, autonomous
bene	good	benevolent, benefactor
bio	life	biology, biography
chron	time	chronology
circum	around	circumvent, circumference
dyna	power	dynasty, dynamite
fort	strength	fortuitous, fortress
graph	writing	graphic
hetero	different	heterogeneous
homo	same	homonym, homogenous
hypo	below, beneath	hypothermia
morph	shape; form	morphology
mort	death	mortal, mortician
multi	many	multimedia, multiplication
nym	name	antonym, synonym
phobia	fear	claustrophobia
port	carry	transport
pseudo	false	pseudoscience, pseudonym
scope	viewing instrument	telescope, microscope
techno	art; science; skill	technology, techno
therm	heat	thermometer, thermal
trans	across	transatlantic, transmit
under	too little	underestimate

Analogies

Verbal Analogies

The analogy section is designed to test your knowledge of word definitions as well as their relationship to one another and as a pair to another pair of words. Making analogies is part of an important cognitive process that acts as the basis of metaphor and association. Analogical thinking is used in problem solving, creative thinking, argumentation, invention, communication, and memory, among other intellectual operations. Verbal analogies are a way to determine associations between objects, their signifying words, and each word's specific connotations. Learning verbal analogies is valued among many standardized exams because of its usefulness in language and learning, especially among figurative language such as metaphors, similes, and allegories.

Question Format

Below we will look at the different types of analogies, but it's also helpful to know what format the analogy section uses and how best to answer the questions. The question will give you either two words or three words to work with. Let's start with an example with only two words:

Cat is to **mammal** as

a. **shoe** is to **foot**.
b. **kitten** is to **feline**.
c. **dolphin** is to **amphibian**.
d. **lizard** is to **reptile.**
e. **lamp** is to **bedroom**.

This type of question requires you to carefully study the relationship between the first pair of words. The first pair of words have the relationship of **category and type.** We see that in the world of mammals, a *cat* is a **type** of mammal, or can be considered a mammal. The word "cat" falls within the classification of "mammal." Now when we look at the answer choices, we have to determine that the first word falls within an appropriate category of the second word. "Shoe" is not a category of "foot," so Choice *A* is incorrect. Choice *B*'s "kitten" and "feline" are too close to each other in meaning. Choice *C* is incorrect because a dolphin does not fall under the class of "amphibian." Rather, it is a mammal. Choice *D* looks like the correct answer. "Lizard" is a type of "reptile" and falls under the category of "reptile." This pair of words has the same relationship as the original pair. Finally, Choice *E* is incorrect. A lamp may reside in the bedroom, but the association is too generalized for this question.

Now let's look at a question with three words to start off with:

Heat is to **scorching** as **cold** is to

a. burning.
b. freezing.
c. melting.
d. ice.
e. Alaska.

In the questions where we are given three terms, we still need to identify the relationship between the first pair of words. Here, we see that this type of analogy is a **degree of intensity**. We are given the word "hot," and we know that "scorching" means "really, really hot." Let's look at what cold's intensity looks like. Choices *A* and *C*, burning and melting, are the opposite of an intensity of cold, so these are incorrect. Choice *B* looks like a good answer. "Freezing" is an intensity of cold, so let's mark this as correct. Choice *D*, ice, is an object that is really cold, and Choice *E*, Alaska, is a place that is really cold. These could be in the running. However, to narrow down between "freezing," "ice," and "Alaska," it's important to look at the closest analogy to the word "scorching." The word is an adjective, not a noun or proper noun, as ice and Alaska are. Since "scorching" and "freezing" both occupy the same part of speech, let's choose "freezing," Choice *B*, as the correct answer.

Types of Analogies

In its most basic form, an analogy compares two different things. An analogy is a pair of words that parallels the situation or relationship given in another pair of words. The table below give examples of the different types of analogies you may see:

Type of Analogy	Relationship	Example
Synonym	The pair of words are alike in meaning.	**Happy** is to **joyous** as **sad** is to **somber**.
Antonym	The pair of words are opposite in meaning.	**Lucky** is to **unfortunate** as **victorious** is to **defeated**.
Part to Whole	One word stands for a whole, and the other word stands as a part to that whole.	**Chapter** is to **novel** as **pupil** is to **eye**.
Category/Type	One thing belongs in a category of another thing.	**Screwdriver** is to **tool** as **apartment** is to **dwelling**.
Object to Function	A pair depicting a tool and the use of that tool.	**Shovel** is to **dig** as **oven** is to **bake**.
Degree of Intensity	The pair of words shows a difference in degree.	**Funny** is to **hysterical** as **interest** is to **adoration**.
Cause and Effect	The pair shows that one word is created by the other word.	**Hard work** is to **success** as **privilege** is to **comfort**.
Symbol and Representation	A word and its representation in the context of a culture.	**Rose** is to **love** as **flag** is to **patriotism**.
Performer to Related Action	A person and their related action.	**Professor** is to **teach** as **doctor** is to **heal**.

Practice Questions

Synonyms

For each of the following questions, choose the one word whose meaning is MOST similar to the word printed in capital letters.

1. WARY
    a. Religious
    b. Adventurous
    c. Tired
    d. Negligent
    e. Cautious

2. PROXIMITY
    a. Estimate
    b. Delicate
    c. Precarious
    d. Splendor
    e. Closeness

3. PARSIMONIOUS
    a. Lavish
    b. Harmonious
    c. Miserly
    d. Careless
    e. Generous

4. PROPRIETY
    a. Ownership
    b. Appropriateness
    c. Patented
    d. Abstinence
    e. Sobriety

5. BOON
    a. Cacophony
    b. Hopeful
    c. Benefit
    d. Squall
    e. Omen

6. STRIFE
    a. Plague
    b. Industrial
    c. Picketing
    d. Eliminate
    e. Conflict

7. QUALM
   a. Calm
   b. Uneasiness
   c. Assertion
   d. Pacify
   e. Victory

8. VALOR
   a. Rare
   b. Coveted
   c. Leadership
   d. Bravery
   e. Honorable

9. ZEAL
   a. Craziness
   b. Resistance
   c. Fervor
   d. Opposition
   e. Apprehension

10. EXTOL
   a. Glorify
   b. Demonize
   c. Chide
   d. Admonish
   e. Criticize

11. REPROACH
   a. Locate
   b. Blame
   c. Concede
   d. Orate
   e. Honor

12. MILIEU
   a. Bacterial
   b. Damp
   c. Ancient
   d. Environment
   e. Uncertain

13. GUILE
   a. Masculine
   b. Stubborn
   c. Naïve
   d. Gullible
   e. Deception

14. ASSENT
    a. Acquiesce
    b. Climb
    c. Assert
    d. Demand
    e. Heighten

15. DEARTH
    a. Grounded
    b. Scarcity
    c. Lethal
    d. Risky
    e. Hearty

16. CONSPICUOUS
    a. Scheme
    b. Obvious
    c. Secretive
    d. Ballistic
    e. Paranoid

17. ONEROUS
    a. Responsible
    b. Generous
    c. Hateful
    d. Burdensome
    e. Wealthy

18. BANAL
    a. Inane
    b. Novel
    c. Painful
    d. Complimentary
    e. Inspired

19. CONTRITE
    a. Tidy
    b. Unrealistic
    c. Contrived
    d. Corrupt
    e. Remorseful

20. MOLLIFY
    a. Pacify
    b. Blend
    c. Negate
    d. Amass
    e. Emote

21. CAPRICIOUS
    a. Skillful
    b. Sanguine
    c. Chaotic
    d. Fickle
    e. Agreeable

22. PALTRY
    a. Appealing
    b. Worthy
    c. Trivial
    d. Fancy
    e. Disgusting

23. SHIRK
    a. Counsel
    b. Evade
    c. Diminish
    d. Sharp
    e. Annoy

24. ASSUAGE
    a. Irritate
    b. Persuade
    c. Argue
    d. Redirect
    e. Soothe

25. TACIT
    a. Unspoken
    b. Shortened
    c. Tenuous
    d. Regal
    e. Timid

26. ADVOCATE
    a. Entice
    b. Brandish
    c. Decline
    d. Support
    e. Align

27. CORROBORATE
    a. Verify
    b. Deny
    c. Forget
    d. Claim
    e. Create

28. SHEER
    a. Opaque
    b. Muddy
    c. Brave
    d. Depressed
    e. Translucent

29. VOW
    a. Liberate
    b. Worship
    c. Lie
    d. Break
    e. Promise

30. INCITE
    a. Calm
    b. Provoke
    c. Smell
    d. Repent
    e. Weaken

Analogies

For each of the following questions, choose the option that best completes the sentence.

1. **Chapter** is to **book** as
    a. Book is to story.
    b. Fable is to myth.
    c. Paragraph is to essay.
    d. Dialogue is to play.
    e. Story is to tale.

2. **Dress** is to **garment** as
    a. Diesel is to fuel.
    b. Month is to year.
    c. Suit is to tie.
    d. Clothing is to wardrobe.
    e. Coat is to winter.

3. **Car** is to **garage** as **plane** is to
    a. Sky
    b. Passenger.
    c. Airport.
    d. Runway.
    e. Hanger.

4. **Trickle** is to **gush** as
    a. Bleed is to cut.
    b. Rain is to snow.
    c. Tepid is to scorching.
    d. Sob is to sniffle.
    e. Ocean is to river.

5. **Acne** is to **dermatologist** as **cataract** is to
    a. Psychologist.
    b. Ophthalmologist.
    c. Otolaryngologist
    d. Otologist.
    e. Orthopedist.

6. **Jump** is to **surprise** as
    a. Run is to walk.
    b. Spook is to scare.
    c. Chuckle is to joke.
    d. Hop is to bunny.
    e. Sadness is to cry.

7. **Knife** is to **slice** as **fork** is to
    a. Cut.
    b. Spear
    c. Mouth.
    d. Spoon.
    e. Eat.

8. **Eat** is to **ate** as **spin** is to
    a. Thread.
    b. Spinned.
    c. Spinning.
    d. Spun.
    e. Spins.

9. **Obviate** is to **preclude** as
    a. Exclude is to include.
    b. Conceal is to avert.
    c. Pontificate is to ponder.
    d. Appease is to placate.
    e. Ostentatious is to poignant.

10. **Acre** is to **area** as **fathom** is to
    a. Depth.
    b. Angle degree.
    c. Wind speed.
    d. Ship.
    e. Width.

11. **Green** is to **blue** as **orange** is to
    a. Red.
    b. Yellow.
    c. Purple.
    d. Green.
    e. Blue.

12. **Peel** is to **orange** as
    a. Fur is to bear.
    b. Shed is to snake.
    c. Peel is to sunburn.
    d. Shell is to coconut.
    e. Fuzz is to peach.

13. **Sycophant** is to **flattery** as **raconteur** is to
    a. Heritage.
    b. Philosophies.
    c. Idioms.
    d. Artwork.
    e. Anecdotes.

14. **Viscous** is to **runny** as
    a. Somber is to merciful.
    b. Vacuous is to nostalgic.
    c. Plunder is to laud.
    d. River is to stream.
    e. Obscure is to unequivocal.

15. **Nadir** is to **zenith** as **valley** is to
    a. Depression.
    b. Pinnacle.
    c. Climb.
    d. Rise
    e. Slope.

16. **Pundit** is to **expertise** as **scholar** is to
    a. Study.
    b. Learning.
    c. Novice.
    d. Teacher.
    e. Erudition.

17. **Blueprint** is to **architect** as
    a. Stethoscope is to doctor.
    b. Model is to train.
    c. Lathe is to craftsman.
    d. Outline is to drawing.
    e. Score is to composer.

18. **Evidence** is to **detective** as **gold** is to
    a. Jeweler.
    b. Pan.
    c. Prospector.
    d. Magnet.
    e. Archeologist.

19. **Odometer** is to **distance** as **caliper** is to
    a. Pressure.
    b. Thickness.
    c. Wind.
    d. Brake.
    e. Body.

20. **Radish** is to **vegetable** as
    a. Garbanzo is to legume.
    b. Pineapple is to berry.
    c. Lettuce is to spinach.
    d. Cucumber is to salad.
    e. Citrus is to fruit.

21. **Adore** is to **appreciate** as **loathe** is to
    a. Detest.
    b. Hate.
    c. Appreciate.
    d. Dislike.
    e. Fear.

22. **Thanksgiving** is to **November** as
    a. Summer is to vacation.
    b. Easter is to spring.
    c. Labor Day is to January.
    d. Holiday is to celebration.
    e. Christmas is to December.

23. **Alligator** is to **reptile** as **elephant** is to
    a. Mammal.
    b. Animal.
    c. Australia.
    d. Marsupial.
    e. Bear.

24. **Nylon** is to **parachute** as
    a. Plexiglass is to glass.
    b. Neoprene is to wetsuit.
    c. Sweater is to wool.
    d. Wood is to bark.
    e. Ice is to cube.

25. **Painter** is to **easel** as **weaver** is to
   a. Pattern.
   b. Tapestry.
   c. Yarn.
   d. Needle.
   e. Loom.

26. **Coarse** is to **rough** as
   a. Smooth is to grainy.
   b. Butter is to sandpaper.
   c. Funny is to amusing.
   d. Hot is to ocean.
   e. Dinner is to breakfast.

27. **Fluctuate** is to **persist** as
   a. Apple is to orange.
   b. Guitar is to instrument.
   c. Tail is to cat.
   d. Tremble is to shake.
   e. Happy is to sad.

28. **Engine** is to **car** as
   a. Lemon is to fruit.
   b. Branch is to tree.
   c. Brave is to courageous.
   d. Rubber is to cement.
   e. Silverware is to cutlery.

29. **Lotion** is to **hydrate** as
   a. Plant is to flower.
   b. Angry is to irate.
   c. Ear is to head.
   d. Pot is to boil.
   e. Fumigate is to vaporize.

30. **Green** is to **go** as
   a. Hurricane is to tornado.
   b. Yellow is to color.
   c. Crispy is to chewy.
   d. Dove is to peace.
   e. Salamander is to snake.

Answer Explanations

Synonyms

1. E: Someone who is *wary* is overly cautious or apprehensive. This word is often used in the context of being watchful or on guard about a potential danger. For example, darkening clouds and white caps on the waves may make a seaman wary against setting sail.

2. E: *Proximity* is defined as closeness, or the state or quality of being near in place, time, or relation.

3. C: As an adjective, *parsimonious* means frugal to the point of being stingy, or very unwilling to spend money, which is similar to being miserly.

4. B: The noun *propriety* means suitability or appropriateness to the given circumstances or purpose. It can mean conformity to accepted standards, particularly as they relate to good behavior or manners. In this way, *propriety* can be considered to mean the state or quality of being proper. For example, beyond simply upholding the law, Americans typically expect their president to act with propriety.

5. C: A *boon* is a benefit or blessing, often considered to be timely. It is something to be thankful for. For example, a new tax benefit enacted for first-time homeowners would be a boon to a family who just closed on a house. In an alternative usage, it can be a favor or a benefit given upon request.

6. E: *Strife* is a noun that is defined as bitter or vigorous discord, conflict, or dissension. It can mean a fight or struggle, or other act of contention. For example, antagonistic political interest groups vying for local support may be at strife.

7. B: *Qualm* is a noun that means a feeling of apprehension or uneasiness, often brought on suddenly. A girl who is just learning to ride a bike may have qualms about getting back on the saddle after taking a bad fall. It may also refer to an uneasy feeling related to one's conscience as it pertains to his or her actions. For example, a man with poor morals may have no qualms about lying on his tax return.

8. D: *Valor* is bravery or courage when facing a formidable danger. It often relates to strength of mind or spirit during battle, or acting heroically in such situations.

9. C: *Zeal* is eagerness, fervor, or ardent desire in the pursuit of something. For example, a competitive collegiate baseball player's zeal to succeed in his sport may compromise his academic performance.

10: A: To *extol* is to highly praise, laud, or glorify. People often extol the achievements of their heroes or mentors.

11. B: To *reproach* means to blame, find fault, or severely criticize. It can also be defined as expressing significant disapproval. It is often used in the phrase "beyond reproach" as in, "her violin performance was beyond reproach." In this context, it means her playing was so good that it evaded any possibility of criticism.

12. D: *Milieu* refers to the surroundings or environment.

13. E: *Guile* can be defined as the quality of being cunning or crafty and skilled in deception. Someone may use guile to trick or deceive someone, like to get money from them, or otherwise dupe them.

14. A: As a verb, *assent* means to express agreement, give consent, or acquiesce. A job candidate might assent to an interviewer's request to perform a background check. As a noun, it means an agreement, acceptance, or acquiescence.

15. B: *Dearth* means a lack of something or a scarcity or shortage. For example, a local library might have a dearth of information pertaining to an esoteric topic.

16. B: *Conspicuous* means to be visually or mentally obvious. Something conspicuous stands out, is clearly visible, or may attract attention.

17. D: *Onerous* most closely means burdensome or troublesome. It usually is used to describe a task or obligation that may impose a hardship or burden, often which may be perceived to outweigh its benefits.

18: A: Something *banal* lacks originality, and may be boring and trite. For example, a banal compliment is likely to be a common platitude. Like something that is inane, a banal compliment might be meaningless and lack a convincing quality or significance.

19. E: *Contrite* means to feel or express remorse, or to be regretful and interested in repenting. The noun *contrition* refers to severe remorse or penitence.

20. A: *Mollify* means to sooth, pacify, or appease. It usually is used to refer to reducing the anger or softening the feelings or temper of another person, or otherwise calm them down. For example, a customer service associate may need to mollify an irate customer who is furious about the defect in his or her purchase.

21. D: Of the provided choices, *fickle* is closest in meaning to *capricious*. A person who is capricious tends to display erratic or unpredictable behavior, which is similar to fickle, which is also likely to change spontaneously or behave erratically.

22. C: Although *paltry* often is used as an adjective to describe a very small or meager amount (of money, in particular), it can also mean something trivial or insignificant.

23. B: The word *shirk* means to evade, and is often used in the context of shirking a responsibility, duty, or work.

24. E: *Assuage* most nearly means to soothe or comfort, as in to assuage one's fears. It can also mean to lessen or make less severe, or to relieve. For example, an ice pack on a swollen knee may assuage the pain.

25. A: Something that is *tacit* is usually unspoken but implied. Tacit approval, for example, occurs when agreement or approval is understood without explicitly stating it.

26. D: To *advocate* means to *support* something or someone. Advocate here is presented as a verb; we see the verb *support* in the answer choices, which is the closest in meaning to *advocate*.

27. A: To *corroborate* is a verb that means to verify or give support to. *Verify* is the best choice here.

28. E: *Sheer* is an adjective that means see through or thin. The word most closely related to *sheer* is *translucent*, which also means see through. If something is *sheer* or *translucent*, you are able to see through it.

29. E: A *vow* is a solemn pledge, oath, or promise made to someone. The word that is the closest synonym to the word *vow* is *promise*, Choice *E*.

30. B: The word *incite* means to encourage or stir up disruptive behavior. A synonym for the word *incite* is *provoke*, which means to stimulate or stir up emotion in someone.

Analogies

1. C: This is a part/whole analogy. A chapter a section, or portion, of a book. All of the choices relate to literary topics, but the best option is paragraph is to essay. A paragraph is a section, or building block of an essay in much the same way that a chapter is in a book. The word pairs in Choices *A, B,* and *E* are best described as near synonyms, but not necessarily parts of one another. Choice *D,* dialogue is to play, does include more of a part to the whole relationship, but dialogue is the way the story is conveyed in a play (like sentences in a book). A better matching analogy to chapter is to book would be scene or act is to play, since plays are divided into scenes or larger acts.

2. A: This is a type of category analogy. A dress is a type of garment. Garment is the broad category, and dress is the specific example used. Diesel is a type of fuel, so it holds the same relationship. Month is not a type of year; it's part of the year. A suit is not a type of tie, but it might be worn with a tie. Clothing is not a type of wardrobe; it is stored in a wardrobe. Lastly, a coat isn't a type of winter; it is a garment worn in the winter.

3. E: This is a provider/provision analogy. The analogy focuses on where the mode of transportation is stored or housed when not in use. A garage is where a car is sheltered and kept when not in use, much like a hanger for an airplane. Choice *C,* airport, might be an appealing choice, but an airport doesn't have as precise of a relationship to a plane as does a garage to a car. An airport might be where planes are located before use or where one might see a lot of planes, like a parking lot for a car.

4. C: This is an intensity analogy, or possibly an antonym analogy. The prevailing connection is somewhat opposite meanings, or certainly opposite intensities. Fluid that is trickling is barely moving or of low volume, while gushing fluid is moving fast and often in a larger volume. The only choices that are related by intensity or are antonyms are Choices *C* and *D,* but *D* reverses the relationship. Sob is a more significant cry versus a small sniffle. Tepid is lukewarm or slightly warm, while scorching is very hot.

5. B: This is a provider/provision analogy. The connection is a condition treated by a certain medical professional. Acne may be treated by a dermatologist, a skin doctor. A cataract, an eye condition, is treated by an ophthalmologist. A psychologist treats psychological or mental conditions. An otolaryngologist is an ear, nose, and throat doctor, so one would not treat a cataract. Similarly, an otologist is an ear doctor, and an orthopedist treats skeletal issues.

6. C: This is a cause and effect analogy. The common thread is a physical reaction (effect) to a cause. Someone might jump from a surprise. Choice *C,* maintains this relationship because someone might chuckle in response to a joke. Choice *E* does contain a type of physical reaction effect (crying) and a cause (sadness), but the relationship is reversed. Moreover, "sadness" isn't really an event or occurrence the same way that a surprise or joke is. Sadness is an emotion.

7. B: This is tool/use analogy. The connection is a simple action that the tool (in this case, a utensil) is used for. The question is made easier by providing the next tool (utensil). A knife is used to slice things (bread, apple, etc.). A fork can be used to spear pieces of food, so the pieces can be lifted to the mouth for eating. Therefore, Choice *B* is the best option. A fork is not used for cutting; that is the function of a knife.

Therefore, Choice *A* is incorrect. Choices *C* and *D* are incorrect because they aren't actions that the fork is used for. Choice *E*, eat, is technically an action the fork is used for, although it is not as precise and similar to spear, making Choice *B* a better answer.

8. D: This is a grammatical analogy. It is comparing a verb in the simple present tense with the same verb in the simple past tense. *Ate* is the past tense of *eat*, and *spun* is the past tense of *spin*.

9. D: This is a synonyms analogy, which matches terms based on their similar meanings. *Obviate* is a verb that can mean to prevent or avoid, or to render something unnecessary. It is usually used with an object, one that is often a difficulty or disadvantage. For example, wearing a helmet while cycling can obviate the risk of a skull injury should a fall occur. *Preclude* is also a verb. It means to prevent something from occurring or existing. For example, a thunderstorm may preclude a picnic. It can also mean to exclude from something. For example, an inability to get wet after surgery would preclude the patient from swimming. Of the answer choices provided, Choice *D*, appease and placate, are the only other synonyms. Like the terms in the prompt, these words are both verbs. *Placate* means to pacify, calm, or make a person less angry or upset. Similarly, *appease* means to pacify or calm someone by fulfilling their demands.

10. A: This can be considered a type of characteristic analogy because it is matching a unit of measure with an example of what that unit measures. An *acre* is an example of a unit of measure for area. Plots of land can be measured in acres, and this value gives information about how much area of land that plot occupies. A *fathom* is an example of a unit to measure depth. It is typically used to refer to water depth.

11. A: This is a characteristic analogy. The connection lies in what color results when yellow is added to the primary color. Green is the secondary color that is formed when yellow is added to blue, just as orange is the secondary color created when yellow is added to red.

12. D: This is a parts/whole analogy. The connection is the outer removable, inedible layer of the fruit. The best choice is *D*, shell is to coconut, because the shell is also the outer, inedible part of the coconut. While peaches do have fuzz, the fuzz is an edible portion of the skin. Therefore, Choice *E* is not as closely related to the question stem as the pair in Choice *D*.

13. E: This is a provider/provision analogy. The connection is a type of person or quality of a person and what that personality trait provides (the output, as a noun). A *sycophant* is someone who dishes out a lot of flattery, or insincere praise, often to better his or her situation. A *raconteur* dishes out or regales people with anecdotes and stories in an interesting way.

14. E: This is an antonyms analogy that matches adjectives with opposite meanings. A *viscous* fluid is thick and slowly moving, while a runny one is thin and flows freely. The only answer choice that is also a pair of adjectives that are antonyms is Choice *E*. When used as an adjective, *obscure* refers to something inconspicuous, unnoticeable, or ambiguous. It usually is used to refer to something with an unclear meaning or hard to understand, such as obscure intentions or the use of obscure language. *Unequivocal* is an adjective that means essentially the exact opposite—unambiguous or leaving no doubt.

15. B: This is another antonyms analogy. *Zenith* and *nadir* are astronomical nouns with opposite meanings that have also been adopted into conversational (non-technical) English. In an astronomical sense, the *zenith* is the point in the celestial sphere or sky that lies directly above the observer, while the *nadir* is the point directly below the observer. These terms have been incorporated into common language to mean the very top or culminating point of something (the *zenith*), and the very bottom, lowest, or worst (the *nadir*). For example, the *zenith* of triathlete's athletic career might be winning the Ironman World Championships, while the *nadir* might be crashing his or her bike during a race and fracturing a bone. In this question, the term *valley* is provided for the next pair. Like *nadir,* a *valley* is a very low point. The correct answer will then be a high point, which is best captured by Choice *B, pinnacle.*

16. E: This analogy matches a characteristic or type of person with the quality that type of person possesses. A *pundit* is an expert in a certain field or particular subject. A pundit has *expertise.* A scholar has learned or gained knowledge in the areas he or she has studied. They have *erudition.*

17. E: This analogy matches an occupation with what someone in that job creates and uses as a plan for their work. An architect creates a blueprint to be a rendering of the plan for the structures that builders will use to erect the building, much like a composer creates a score that musicians will follow to play the music. The other answer choices do not maintain this same relationship.

18. C: This is a type of tool/user analogy. Rather than simply being a "tool" used by the user, it pairs the user with what they seek. A detective searches for evidence or clues, while a prospector searches for gold.

19. B: This is a type of tool/use analogy. It matches a tool and what it is used to measure. An odometer is used to measure distance traveled in a car. A caliper is tool used to measure thickness. For example, personal trainers use skinfold calipers to measure the thickness of various folds of skin to estimate body fat.

20. A: This is a category analogy. The connection is the way that food is classified. A radish is a type of vegetable and a garbanzo bean (chickpea) is a type of legume. Pineapples are not a type of berry, lettuce is not a type of spinach, and cucumber is not a type of salad. Therefore, Choices *B, C,* and *D* are incorrect. Choice *E* may look appealing because citrus is a type of fruit, but "citrus" isn't a specific fruit. For the analogy to hold, it would need to be a specific citrus fruit, like lemon.

21. D: This is an intensity analogy. *Adore* is a stronger version of *appreciate. Loathe* is a stronger version of *dislike. Detest* and *hate,* Choices *A* and *B,* are more synonymous with *loathe* (rather than a different intensity), so they are not the best choice. Choice *C* is an antonym, and Choice *E* is unrelated.

22. E: This analogy makes use of the temporal relationship between an event (a holiday) and the calendar month that it occurs. Thanksgiving is a holiday in November, like Christmas is a holiday in December. Choice *C* is incorrect because Labor Day is in September. The other choices do not relate a specific holiday to the particular month in which they occur.

23. A: This is a category analogy. The common thread is the classification of the animal. An alligator is a type of reptile, just as an elephant is a type of mammal.

24. B: This is a source/comprised of analogy that focuses on pairing a raw material with an item that's created from it. Nylon is used to make parachutes just as neoprene is used to make wetsuits. Although sweaters are made from wool, the relationship is reversed. Test takers should remember to be careful about maintaining the same order of the words in the relationship when solving analogies; otherwise, the meaning is changed.

25. E: This is a tool/user analogy. The connection is the type of tool the artist uses to hold and create their work. An easel holds the paper or canvas that a painter uses to create a painting much like a loom is the apparatus used to hold and weave yarns into a blanket or other tapestry.

26. C: This is a synonyms analogy, which matches terms based on similar meanings. *Coarse* and *rough* mean the same thing, so they are synonyms. Now let's look at another pair in the answer choices that share a similar meaning. Choice *C* is the best choice we have. *Funny* and *amusing* are synonyms, because they both mean that someone is acting comical. The other answer choices do not share the same meaning with each other.

27. E: This is an antonyms analogy, which matches terms that have opposite meanings. *Fluctuate* means to sway, waver, or change, so the word *persist*, which means to stay or remain stable, is the opposite of *fluctuate*. The word pair in the answer choices that have opposite meanings are *happy* and *sad*, Choice *E*.

28. B: This is a part to whole analogy. An *engine* is a part that *makes up* a car. Let's look at two words that represent a part/whole analogy. Choice *B* is the correct answer, because a *branch* is part of a *tree*.

29. D: This analogy is an object to function. *Lotion* is an object used to moisturize or *hydrate*. Let's look at another object that shows its use through the answer choices. A *pot* is an object used to *boil*, so Choice *D* is the best choice.

30. D: This is a symbol/representation analogy. *Green* represents *go*. When people see a green light, especially driving on the road, it is a signal for them to keep driving. Similarly, a *dove* represents *peace*. In literature and in history, whenever there is a dove, it is meant to represent peace or calm.

Writing

Conventions of Standard English Spelling

Homophones
Homophones are words that have different meanings and spellings, but sound the same. These can be confusing for English Language Learners (ELLs) and beginning students, but even native English-speaking adults can find them problematic unless informed by context. Whereas listeners must rely entirely on context to differentiate spoken homophone meanings, readers with good spelling knowledge have a distinct advantage since homophones are spelled differently. For instance, *their* means belonging to them; *there* indicates location; and *they're* is a contraction of *they are*; despite different meanings, they all sound the same. *Lacks* can be a plural noun or a present-tense, third-person singular verb; either way it refers to absence—*deficiencies* as a plural noun, and *is deficient in* as a verb. But *lax* is an adjective that means loose, slack, relaxed, uncontrolled, or negligent. These two spellings, derivations, and meanings are completely different. With speech, listeners cannot know spelling and must use context; but with print, readers with spelling knowledge can differentiate them with or without context.

Homonyms, Homophones, and Homographs
As just mentioned, **homophones** are words that sound the same in speech, but have different spellings and meanings. For example, *to, too,* and *two* all sound alike, but have three different spellings and meanings. Homophones with different spellings are also called **heterographs. Homographs** are words that are spelled identically, but have different meanings. If they also have different pronunciations, they are **heteronyms.** For instance, *tear* pronounced one way means a drop of liquid formed by the eye; pronounced another way, it means to rip. Homophones that are also homographs are **homonyms.** For example, *bark* can mean the outside of a tree or a dog's vocalization; both meanings have the same spelling. *Stalk* can mean a plant stem or to pursue and/or harass somebody; these are spelled and pronounced the same. *Rose* can mean a flower or the past tense of *rise.* Many non-linguists confuse things by using "homonym" to mean sets of words that are homophones but not homographs, and also those that are homographs but not homophones.

The word *row* can mean to use oars to propel a boat; a linear arrangement of objects or print; or an argument. It is pronounced the same with the first two meanings, but differently with the third. Because it is spelled identically regardless, all three meanings are homographs. However, the two meanings pronounced the same are homophones, whereas the one with the different pronunciation is a heteronym. By contrast, the word *read* means to peruse language, whereas the word *reed* refers to a marsh plant. Because these are pronounced the same way, they are homophones; because they are spelled differently, they are heterographs. Homonyms are both homophones and **homographs**—words that are pronounced and spelled identically, but with different meanings. One distinction between homonyms is of those with separate, unrelated etymologies, called "true" homonyms, e.g. *skate* meaning a fish or *skate* meaning to glide over ice/water. Those with common origins are called polysemes or polysemous homonyms, e.g. the *mouth* of an animal/human or of a river.

Irregular Plurals
One type of irregular English plural involves words that are spelled the same whether they are singular or plural. These include *deer, fish, salmon, trout, sheep, moose, offspring, species, aircraft,* etc. The spelling rule for making these words plural is simple: they do not change. Another type of irregular English plurals does change from singular to plural form, but it does not take regular English *–s* or *–es* endings. Their irregular plural endings are largely derived from grammatical and spelling conventions in the other

languages of their origins, like Latin, German, and vowel shifts and other linguistic mutations. Some examples of these words and their irregular plurals include *child* and *children; die* and *dice; foot* and *feet; goose* and *geese; louse* and *lice; man* and *men; mouse* and *mice; ox* and *oxen; person* and *people; tooth* and *teeth;* and *woman* and *women.*

Contractions
Contractions are formed by joining two words together, omitting one or more letters from one of the component words, and replacing the omitted letter/s with an apostrophe. An obvious yet often forgotten rule for spelling contractions is to place the apostrophe where the letters were omitted; for example, spelling errors like *did'nt* for *didn't. Didn't* is a contraction of *did not.* Therefore, the apostrophe replaces the "o" that is omitted from the "not" component. Another common error is confusing contractions with possessives because both include apostrophes, e.g. spelling the possessive *its* as "it's," which is a contraction of "it is"; spelling the possessive *their* as "they're," a contraction of "they are"; spelling the possessive *whose* as "who's," a contraction of "who is"; or spelling the possessive *your* as "you're," a contraction of "you are."

Frequently Misspelled Words
One source of spelling errors is not knowing whether to drop the final letter *e* from a word when its form is changed by adding an ending to indicate the past tense or progressive participle of a verb, converting an adjective to an adverb, a noun to an adjective, etc. Some words retain the final *e* when another syllable is added; others lose it. For example, *true* becomes *truly; argue* becomes *arguing; come* becomes *coming; write* becomes *writing;* and *judge* becomes *judging.* In these examples, the final *e* is dropped before adding the ending. But *severe* becomes *severely; complete* becomes *completely; sincere* becomes *sincerely; argue* becomes *argued;* and *care* becomes *careful.* In these instances, the final *e* is retained before adding the ending. Note that some words, like *argue* in these examples, drops the final *e* when the *–ing* ending is added to indicate the participial form; but the regular past tense ending of *–ed* makes it *argued,* in effect replacing the final *e* so that *arguing* is spelled without an *e* but *argued* is spelled with one.

Some English words contain the vowel combination of *ei,* while some contain the reverse combination of *ie.* Many people confuse these. Some examples include these:

> *ceiling, conceive, leisure, receive, weird, their, either, foreign, sovereign, neither, neighbors, seize, forfeit, counterfeit, height, weight, protein,* and *freight*

Words with *ie* include *piece, believe, chief, field, friend, grief, relief, mischief, siege, niece, priest, fierce, pierce, achieve, retrieve, hygiene, science,* and *diesel.* A rule that also functions as a mnemonic device is "I before E except after C, or when sounded like A as in 'neighbor' or 'weigh'." However, it is obvious from the list above that many exceptions exist.

Many people often misspell certain words by confusing whether they have the vowel *a, e,* or *i,* frequently in the middle syllable of three-syllable words or beginning the last syllables that sound the same in different words. For example, in the following correctly spelled words, the vowel in boldface is the one people typically get wrong by substituting one or either of the others for it:

> cem**e**tery, quant**i**ties, ben**e**fit, priv**i**lege, unpleas**a**nt, sep**a**rate, independ**e**nt, excell**e**nt, cat**e**gories, indispens**a**ble, and irrelev**a**nt

The words with final syllables that sound the same when spoken but are spelled differently include *unpleasant, independent, excellent,* and *irrelevant.* Another source of misspelling is whether or not to

225

double consonants when adding suffixes. For example, we double the last consonant before *–ed* and *–ing* endings in *controlled, beginning, forgetting, admitted, occurred, referred,* and *hopping;* but we do not double the last consonant before the suffix in *shining, poured, sweating, loving, hating, smiling,* and *hoping.*

One way in which people misspell certain words frequently is by failing to include letters that are silent. Some letters are articulated when pronounced correctly but elided in some people's speech, which then transfers to their writing. Another source of misspelling is the converse: people add extraneous letters. For example, some people omit the silent *u* in *guarantee,* overlook the first *r* in *surprise,* leave out the *z* in *realize,* fail to double the *m* in *recommend,* leave out the middle *i* from *aspirin,* and exclude the *p* from *temperature.* The converse error, adding extra letters, is common in words like *until* by adding a second *l* at the end; or by inserting a superfluous syllabic *a* or *e* in the middle of *athletic,* reproducing a common mispronunciation.

Conventions of Standard English Punctuation

Rules of Capitalization
The first letter of the first word of any document, and of each new sentence, is capitalized. The first letter of all proper nouns, and names and adjectives derived from proper nouns, should also be capitalized. Here are some examples:

- Grand Canyon
- Pacific Palisades
- Golden Gate Bridge
- Freudian slip
- Shakespearian, Spenserian, or Petrarchan sonnet
- Irish song

Some exceptions are adjectives, originally derived from proper nouns, but through time and usage are no longer capitalized, like *quixotic, herculean,* or *draconian.* Capitals draw attention to specific instances of people, places, and things. Some categories that should be capitalized include the following:

- brand names
- companies
- weekdays
- months
- governmental divisions or agencies
- historical eras
- major historical events
- holidays
- institutions
- famous buildings
- ships and other manmade constructions
- natural and manmade landmarks
- territories
- nicknames
- epithets
- organizations
- planets
- nationalities

- tribes
- religions
- names of religious deities
- roads
- special occasions, like the Cannes Film Festival or the Olympic Games

Exceptions

Related to American government, capitalize the noun *Congress* but not the related adjective *congressional*. Capitalize the noun *U.S. Constitution*, but not the related adjective *constitutional*. Many experts advise leaving the adjectives *federal* and *state* in lowercase, as in federal regulations or state water board, and only capitalizing these when they are parts of official titles or names, like Federal Communications Commission or State Water Resources Control Board. While the names of the other planets in the solar system are capitalized as names, *Earth* is more often capitalized only when being described specifically as a planet, like Earth's orbit, but lowercase otherwise since it is used not only as a proper noun but also to mean *land, ground, soil*, etc.

Names of animal species or breeds are not capitalized unless they include a proper noun. Then, only the proper noun is capitalized. *Antelope, black bear*, and *yellow-bellied sapsucker* are not capitalized. However, Bengal tiger, German shepherd, Australian shepherd, French poodle, and Russian blue cat are capitalized.

Other than planets, celestial bodies like the *sun, moon,* and *stars* are not capitalized. Medical conditions like *tuberculosis* or *diabetes* are lowercase; again, exceptions are proper nouns, like Epstein-Barr syndrome, Alzheimer's disease, and Down syndrome. Seasons and related terms like winter solstice or autumnal equinox are lowercase. Plants, including fruits and vegetables, like *poinsettia, celery*, or *avocados*, are not capitalized unless they include proper names, like Douglas fir, Jerusalem artichoke, Damson plums, or Golden Delicious apples.

Titles and Names

When official titles precede names, they should be capitalized, except when there is a comma between the title and name. But if a title follows or replaces a name, it should not be capitalized. For example, "the president" without a name is not capitalized, as in "The president addressed Congress." But with a name, it is capitalized, like "President Obama addressed Congress." Or, "Chair of the Board Janet Yellen was appointed by President Obama." One exception is that some publishers and writers nevertheless capitalize President, Queen, Pope, etc., when these are not accompanied by names to show respect for these high offices. However, many writers in America object to this practice for violating democratic principles of equality. Occupations before full names are not capitalized, like owner Mark Cuban, director Martin Scorsese, or coach Roger McDowell.

Some universal rules for capitalization in composition titles include capitalizing the following:

- The first and last words of the title
- Forms of the verb to be and all other verbs
- Pronouns
- The word not

Universal rules for NOT capitalizing include the articles *the, a,* or *an;* the conjunctions *and, or,* or *nor,* and the preposition *to,* or *to* as part of the infinitive form of a verb. The exception to all of these is UNLESS any of them is the first or last word in the title, in which case they are capitalized. Other words are subject to differences of opinion and differences among various stylebooks or methods. These include *as, but, if,* and *or,* which some capitalize and others do not. Some authorities say no preposition should ever be capitalized; some say prepositions five or more letters long should be capitalized. The *Associated Press Stylebook* advises capitalizing prepositions longer than three letters (like *about, across,* or *with*).

Ellipses

Ellipses (. . .) signal omitted text when quoting. Some writers also use them to show a thought trailing off, but this should not be overused outside of dialogue. An example of an ellipses would be if someone is quoting a phrase out of a professional source but wants to omit part of the phrase that isn't needed: "Dr. Skim's analysis of pollen inside the body is clearly a myth . . . that speaks to the environmental guilt of our society."

Commas

Commas separate words or phrases in a series of three or more. The Oxford comma is the last comma in a series. Many people omit this last comma, but many times, doing so causes confusion. Here is an example:

> I love my sisters, the Queen of England and Madonna.

This example without the comma implies that the "Queen of England and Madonna" are the speaker's sisters. However, if the speaker was trying to say that they love their sisters, the Queen of England, as well as Madonna, there should be a comma after "Queen of England" to signify this.

Commas also separate two coordinate adjectives ("big, heavy dog") but not cumulative ones, which should be arranged in a particular order for them to make sense ("beautiful ancient ruins").

A comma ends the first of two independent clauses connected by conjunctions. Here is an example:

> I ate a bowl of tomato soup, and I was hungry very shortly after.

Here are some brief rules for commas:

- Commas follow introductory words like *however, furthermore, well, why,* and *actually,* among others.

- Commas go between city and state: Houston, Texas.

- If using a comma between a surname and Jr. or Sr. or a degree like M.D., also follow the whole name with a comma: "Martin Luther King, Jr., wrote that."

- A comma follows a dependent clause beginning a sentence: "Although she was very small, . . ."

- Nonessential modifying words/phrases/clauses are enclosed by commas: "Wendy, who is Peter's sister, closed the window."

- Commas introduce or interrupt direct quotations: "She said, 'I hate him.' 'Why,' I asked, 'do you hate him?'"

Semicolons

Semicolons are used to connect two independent clauses, but should never be used in the place of a comma. They can replace periods between two closely connected sentences: "Call back tomorrow; it can wait until then." When writing items in a series and one or more of them contains internal commas, separate them with semicolons, like the following:

People came from Springfield, Illinois; Alamo, Tennessee; Moscow, Idaho; and other locations.

Hyphens

Here are some rules concerning hyphens:

- Compound adjectives like *state-of-the-art* or *off-campus* are hyphenated.

- Original compound verbs and nouns are often hyphenated, like "throne-sat," "video-gamed," "no-meater."

- Adjectives ending in *–ly* are often hyphenated, like "family-owned" or "friendly-looking."

- "Five years old" is not hyphenated, but singular ages like "five-year-old" are.

- Hyphens can clarify. For example, in "stolen vehicle report," "stolen-vehicle report" clarifies that "stolen" modifies "vehicle," not "report."

- Compound numbers twenty-one through ninety-nine are spelled with hyphens.

- Prefixes before proper nouns/adjectives are hyphenated, like "mid-September" and "trans-Pacific."

Parentheses

Parentheses enclose information such as an aside or more clarifying information: "She ultimately replied (after deliberating for an hour) that she was undecided." They are also used to insert short, in-text definitions or acronyms: "His FBS (fasting blood sugar) was higher than normal." When parenthetical information ends the sentence, the period follows the parentheses: "We received new funds ($25,000)." Only put periods within parentheses if the whole sentence is inside them: "Look at this. (You'll be astonished.)" However, this can also be acceptable as a clause: "Look at this (you'll be astonished)." Although parentheses appear to be part of the sentence subject, they are not, and do not change subject-verb agreement: "Will (and his dog) was there."

Quotation Marks

Quotation marks are typically used when someone is quoting a direct word or phrase someone else writes or says. Additionally, quotation marks should be used for the titles of poems, short stories, songs, articles, chapters, and other shorter works. When quotations include punctuation, periods and commas should *always* be placed inside of the quotation marks.

When a quotation contains another quotation inside of it, the outer quotation should be enclosed in double quotation marks and the inner quotation should be enclosed in single quotation marks. For example: "Timmy was begging, 'Don't go! Don't leave!'" When using both double and single quotation marks, writers will find that many word-processing programs may automatically insert enough space between the single and double quotation marks to be visible for clearer reading. But if this is not the case, the writer should write/type them with enough space between to keep them from looking like three single quotation marks. Additionally, non-standard usages, terms used in an unusual fashion, and technical terms are often clarified by quotation marks. Here are some examples:

My "friend," Dr. Sims, has been micromanaging me again.

This way of extracting oil has been dubbed "fracking."

Apostrophes

One use of the apostrophe is followed by an *s* to indicate possession, like *Mrs. White's home* or *our neighbor's dog*. When using the *'s* after names or nouns that also end in the letter *s*, no single rule applies: some experts advise adding both the apostrophe and the *s*, like "the Jones's house," while others prefer using only the apostrophe and omitting the additional *s*, like "the Jones' house." The wisest expert advice is to pick one formula or the other and then apply it consistently. Newspapers and magazines often use *'s* after common nouns ending with *s*, but add only the apostrophe after proper nouns or names ending with *s*. One common error is to place the apostrophe before a name's final *s* instead of after it: "Ms. Hasting's book" is incorrect if the name is Ms. Hastings.

Plural nouns should not include apostrophes (e.g. "apostrophe's"). Exceptions are to clarify atypical plurals, like verbs used as nouns: "These are the do's and don'ts." Irregular plurals that do not end in *s* always take apostrophe-*s*, not *s*-apostrophe—a common error, as in "childrens' toys," which should be "children's toys." Compound nouns like mother-in-law, when they are singular and possessive, are followed by apostrophe-*s*, like "your mother-in-law's coat." When a compound noun is plural and possessive, the plural is formed before the apostrophe-*s*, like "your sisters-in-laws' coats." When two people named possess the same thing, use apostrophe-*s* after the second name only, like "Dennis and Pam's house."

Sentence Structures

Incomplete Sentences

Four types of incomplete sentences are sentence fragments, run-on sentences, subject-verb and/or pronoun-antecedent disagreement, and non-parallel structure.

Sentence fragments are caused by absent subjects, absent verbs, or dangling/uncompleted dependent clauses. Every sentence must have a subject and a verb to be complete. An example of a fragment is "Raining all night long," because there is no subject present. "It was raining all night long" is one correction. Another example of a sentence fragment is the second part in "Many scientists think in unusual ways. Einstein, for instance." The second phrase is a fragment because it has no verb. One correction is "Many scientists, like Einstein, think in unusual ways." Finally, look for "cliffhanger" words like *if, when, because,* or *although* that introduce dependent clauses, which cannot stand alone without an independent clause. For example, to correct the sentence fragment "If you get home early," add an independent clause: "If you get home early, we can go dancing."

Run-On Sentences

A run-on sentence combines two or more complete sentences without punctuating them correctly or separating them. For example, a run-on sentence caused by a lack of punctuation is the following:

> There is a malfunction in the computer system however there is nobody available right now who knows how to troubleshoot it.

One correction is, "There is a malfunction in the computer system; however, there is nobody available right now who knows how to troubleshoot it." Another is, "There is a malfunction in the computer system. However, there is nobody available right now who knows how to troubleshoot it."

An example of a comma splice of two sentences is the following:

> Jim decided not to take the bus, he walked home.

Replacing the comma with a period or a semicolon corrects this. Commas that try and separate two independent clauses without a contraction are considered comma splices.

Parallel Sentence Structures

Parallel structure in a sentence matches the forms of sentence components. Any sentence containing more than one description or phrase should keep them consistent in wording and form. Readers can easily follow writers' ideas when they are written in parallel structure, making it an important element of correct sentence construction. For example, this sentence lacks parallelism: "Our coach is a skilled manager, a clever strategist, and works hard." The first two phrases are parallel, but the third is not. Correction: "Our coach is a skilled manager, a clever strategist, and a hard worker." Now all three phrases match in form. Here is another example:

> Fred intercepted the ball, escaped tacklers, and a touchdown was scored.

This is also non-parallel. Here is the sentence corrected:

> Fred intercepted the ball, escaped tacklers, and scored a touchdown.

Sentence Fluency

For fluent composition, writers must use a variety of sentence types and structures, and also ensure that they smoothly flow together when they are read. To accomplish this, they must first be able to identify fluent writing when they read it. This includes being able to distinguish among simple, compound, complex, and compound-complex sentences in text; to observe variations among sentence types, lengths, and beginnings; and to notice figurative language and understand how it augments sentence length and imparts musicality. Once students/writers recognize superior fluency, they should revise their own writing to be more readable and fluent. They must be able to apply acquired skills to revisions before being able to apply them to new drafts.

One strategy for revising writing to increase its sentence fluency is flipping sentences. This involves rearranging the word order in a sentence without deleting, changing, or adding any words. For example, the student or other writer who has written the sentence, "We went bicycling on Saturday" can revise it to, "On Saturday, we went bicycling." Another technique is using appositives. An appositive is a phrase or word that renames or identifies another adjacent word or phrase. Writers can revise for sentence fluency by inserting main phrases/words from one shorter sentence into another shorter sentence, combining them into one longer sentence, e.g. from "My cat Peanut is a gray and brown tabby. He loves hunting rats." to "My cat Peanut, a gray and brown tabby, loves hunting rats." Revisions can also connect shorter sentences by using conjunctions and commas and removing repeated words: "Scott likes eggs. Scott is allergic to eggs" becomes "Scott likes eggs, but he is allergic to them."

One technique for revising writing to increase sentence fluency is "padding" short, simple sentences by adding phrases that provide more details specifying why, how, when, and/or where something took place. For example, a writer might have these two simple sentences: "I went to the market. I purchased a cake." To revise these, the writer can add the following informative dependent and independent clauses and prepositional phrases, respectively: "Before my mother woke up, I sneaked out of the house and went to the supermarket. As a birthday surprise, I purchased a cake for her." When revising sentences to make them longer, writers must also punctuate them correctly to change them from simple sentences to compound, complex, or compound-complex sentences.

Skills Writers Can Employ to Increase Fluency
One way writers can increase fluency is by varying the beginnings of sentences. Writers do this by starting most of their sentences with different words and phrases rather than monotonously repeating the same ones across multiple sentences. Another way writers can increase fluency is by varying the lengths of sentences. Since run-on sentences are incorrect, writers make sentences longer by also converting them from simple to compound, complex, and compound-complex sentences. The coordination and subordination involved in these also give the text more variation and interest, hence more fluency. Here are a few more ways writers can increase fluency:

- Varying the transitional language and conjunctions used makes sentences more fluent.
- Writing sentences with a variety of rhythms by using prepositional phrases.
- Varying sentence structure adds fluency.

Use Grammar to Enhance Clarity in Writing

Possessives
Possessive forms indicate possession, i.e. that something belongs to or is owned by someone or something. As such, the most common parts of speech to be used in possessive form are adjectives, nouns, and pronouns. The rule for correctly spelling/punctuating possessive nouns and proper nouns is with -*'s*, like "the woman's briefcase" or "Frank's hat." With possessive adjectives, however, apostrophes are not used: these include *my, your, his, her, its, our,* and *their*, like "my book," "your friend," "his car," "her house," "its contents," "our family," or "their property." Possessive pronouns include *mine, yours, his, hers, its, ours,* and *theirs*. These also have no apostrophes. The difference is that possessive adjectives take direct objects, whereas possessive pronouns replace them. For example, instead of using two possessive adjectives in a row, as in "I forgot my book, so Blanca let me use her book," which reads monotonously, replacing the second one with a possessive pronoun reads better: "I forgot my book, so Blanca let me use hers."

Pronouns
There are three pronoun cases: subjective case, objective case, and possessive case. Pronouns as subjects are pronouns that replace the subject of the sentence, such as *I, you, he, she, it, we, they* and *who*. Pronouns as objects replace the object of the sentence, such as *me, you, him, her, it, us, them*, and *whom*. Pronouns that show possession are *mine, yours, hers, its, ours, theirs*, and *whose*. The following are examples of different pronoun cases:

- Subject pronoun: She ate the cake for her birthday. I saw the movie.
- Object pronoun: You gave me the card last weekend. She gave the picture to him.
- Possessive pronoun: That bracelet you found yesterday is mine. His name was Casey.

Adjectives
Adjectives are descriptive words that modify nouns or pronouns. They may occur before or after the nouns or pronouns they modify in sentences. For example, in "This is a big house," *big* is an adjective modifying or describing the noun *house*. In "This house is big," the adjective is at the end of the sentence rather than preceding the noun it modifies.

A rule of punctuation that applies to adjectives is to separate a series of adjectives with commas. For example, "Their home was a large, rambling, old, white, two-story house." A comma should never separate the last adjective from the noun, though.

Whereas adjectives modify and describe nouns or pronouns, adverbs modify and describe adjectives, verbs, or other adverbs. Adverbs can be thought of as answers to questions in that they describe when, where, how, how often, how much, or to what extent.

Many (but not all) adjectives can be converted to adverbs by adding *–ly*. For example, in "She is a quick learner," *quick* is an adjective modifying *learner*. In "She learns quickly," *quickly* is an adverb modifying *learns*. One exception is *fast*. *Fast* is an adjective in "She is a fast learner." However, *–ly* is never added to the word *fast*; it retains the same form as an adverb in "She learns fast."

Verbs
A verb is a word or phrase that expresses action, feeling, or state of being. Verbs explain what their subject is *doing*. Three different types of verbs used in a sentence are action verbs, linking verbs, and helping verbs.

Action verbs show a physical or mental action. Some examples of action verbs are *play, type, jump, write, examine, study, invent, develop,* and *taste*. The following example uses an action verb:

Kat *imagines* that she is a mermaid in the ocean.

The verb *imagines* explains what Kat is doing: she is imagining being a mermaid.

Linking verbs connect the subject to the predicate without expressing an action. The following sentence shows an example of a linking verb:

The mango *tastes* sweet.

The verb *tastes* is a linking verb. The mango doesn't *do* the tasting, but the word *taste* links the mango to its predicate, sweet. Most linking verbs can also be used as action verbs, such as *smell, taste, look, seem, grow,* and *sound*. Saying something *is* something else is also an example of a linking verb. For example, if we were to say, "Peaches is a dog," the verb *is* would be a linking verb in this sentence, since it links the subject to its predicate.

Helping verbs are verbs that help the main verb in a sentence. Examples of helping verbs are *be, am, is, was, have, has, do, did, can, could, may, might, should,* and *must,* among others. The following are examples of helping verbs:

Jessica *is* planning a trip to Hawaii.

Brenda *does* not like camping.

Xavier *should* go to the dance tonight.

Notice that after each of these helping verbs is the main verb of the sentence: *planning, like,* and *go*. Helping verbs usually show an aspect of time.

Transitional Words and Phrases
In connected writing, some sentences naturally lead to others, whereas in other cases, a new sentence expresses a new idea. We use transitional phrases to connect sentences and the ideas they convey. This makes the writing coherent. Transitional language also guides the reader from one thought to the next. For example, when pointing out an objection to the previous idea, starting a sentence with "However," "But," or "On the other hand" is transitional. When adding another idea or detail, writers use "Also," "In

addition," "Furthermore," "Further," "Moreover," "Not only," etc. Readers have difficulty perceiving connections between ideas without such transitional wording.

Subject-Verb Agreement

Lack of subject-verb agreement is a very common grammatical error. One of the most common instances is when people use a series of nouns as a compound subject with a singular instead of a plural verb. Here is an example:

> Identifying the best books, locating the sellers with the lowest prices, and paying for them *is* difficult

This sentence contains a subject-verb disagreement because it says "is difficult" instead of saying "*are* difficult." Additionally, when a sentence subject is compound, the verb is plural:

> He and his cousins *were* at the reunion.

However, if the conjunction connecting two or more singular nouns or pronouns is "or" or "nor," the verb must be singular to agree:

> That pen or another one like it is in the desk drawer.

If a compound subject includes both a singular noun and a plural one, and they are connected by "or" or "nor," the verb must agree with the subject closest to the verb: "Sally or her sisters go jogging daily"; but "Her sisters or Sally goes jogging daily."

Simply put, singular subjects require singular verbs and plural subjects require plural verbs. A common source of agreement errors is not identifying the sentence subject correctly. For example, people often write sentences incorrectly like, "The group of students *were* complaining about the test." The subject is not the plural "students" but the singular "group." Therefore, the correct sentence should read, "The group of students *was* complaining about the test." The converse also applies, for example, in this incorrect sentence: "The facts in that complicated court case *is* open to question." The subject of the sentence is not the singular "case" but the plural "facts." Hence the sentence would correctly be written: "The facts in that complicated court case *are* open to question." New writers should not be misled by the distance between the subject and verb, especially when another noun with a different number intervenes as in these examples. The verb must agree with the subject, not the noun closest to it.

Pronoun-Antecedent Agreement

Pronouns within a sentence must refer specifically to one noun, known as the **antecedent.** Sometimes, if there are multiple nouns within a sentence, it may be difficult to ascertain which noun belongs to the pronoun. It's important that the pronouns always clearly reference the nouns in the sentence so as not to confuse the reader. Here's an example of an unclear pronoun reference:

> After Catherine cut Libby's hair, David bought her some lunch.

The pronoun in the examples above is *her.* The pronoun could either be referring to *Catherine* or *Libby.* Here are some ways to write the above sentence with a clear pronoun reference:

> After Catherine cut Libby's hair, David bought Libby some lunch.

> David bought Libby some lunch after Catherine cut Libby's hair.

But many times the pronoun will clearly refer to its antecedent, like the following:

> After David cut Catherine's hair, he bought her some lunch.

Formal and Informal Language

Formal language is less personal than informal language. It is more "buttoned-up" and business-like, adhering to proper grammatical rules. It is used in professional or academic contexts, to convey respect or authority. For example, one would use formal language to write an informative or argumentative essay for school or to address a superior. Formal language avoids contractions, slang, colloquialisms, and first-person pronouns. Formal language uses sentences that are usually more complex and often in passive voice. Punctuation can differ as well. For example, exclamation points (!) are used to show strong emotion or can be used as an interjection but should be used sparingly in formal writing situations.

Informal language is often used when communicating with family members, friends, peers, and those known more personally. It is more casual, spontaneous, and forgiving in its conformity to grammatical rules and conventions. Informal language is used for personal emails and correspondence between coworkers or other familial relationships. The tone is more relaxed. In informal writing, slang, contractions, clichés, and the first- and second-person are often used.

Elements of the Writing Process

Skilled writers undergo a series of steps that comprise the writing process. The purpose of adhering to a structured approach to writing is to develop clear, meaningful, coherent work.

The stages are pre-writing or planning, organizing, drafting/writing, revising, and editing. Not every writer will necessarily follow all five stages for every project, but will judiciously employ the crucial components of the stages for most formal or important work. For example, a brief informal response to a short reading passage may not necessitate the need for significant organization after idea generation, but larger assignments and essays will likely mandate use of the full process.

Pre-Writing/Planning
Brainstorming
One of the most important steps in writing is pre-writing. Before drafting an essay or other assignment, it's helpful to think about the topic for a moment or two, in order to gain a more solid understanding of what the task is. Then, spend about five minutes jotting down the immediate ideas that could work for the essay. Brainstorming is a way to get some words on the page and offer a reference for ideas when drafting. Scratch paper is provided for writers to use any pre-writing techniques such as webbing, freewriting, or listing. Some writers prefer using graphic organizers during this phase. The goal is to get ideas out of the mind and onto the page.

Freewriting
Like brainstorming, freewriting is another prewriting activity to help the writer generate ideas. This method involves setting a timer for two or three minutes and writing down all ideas that come to mind about the topic using complete sentences. Once time is up, writers should review the sentences to see what observations have been made and how these ideas might translate into a more unified direction for the topic. Even if sentences lack sense as a whole, freewriting is an excellent way to get ideas onto the page in the very beginning stages of writing. Using complete sentences can make this a bit more challenging than brainstorming, but overall it is a worthwhile exercise, as it may force the writer to come up with more complete thoughts about the topic.

Once the ideas are on the page, it's time for the writer to turn them into a solid plan for the essay. The best ideas from the brainstorming results can then be developed into a more formal outline.

Organizing

Although sometimes it is difficult to get going on the brainstorming or prewriting phase, once ideas start flowing, writers often find that they have amassed too many thoughts, which, if not sorted and weeded, would make for an essay lacking cohesion, direction, and clarity. During the organization stage, writers should examine the generated ideas, hone in on the important ones central to their main idea, and arrange the points in a logical and effective manner. Writers may also determine that some of the ideas generated in the planning process need further elaboration, potentially necessitating the need for research to gather information to fill the gaps.

Once a writer has chosen his or her thesis and main argument, selected the most applicable details and evidence, and eliminated the "clutter," it is time to strategically organize the ideas. This is often accomplished with an outline.

Outlining

An **outline** is a system used to organize writing. When composing essays, outlining is important because it helps writers organize important information in a logical pattern using Roman numerals. Usually, outlines start out with the main ideas and then branch out into subgroups or subsidiary thoughts or subjects. Not only do outlines provide a visual tool for writers to reflect on how events, ideas, evidence, or other key parts of the argument relate to one another, but they can also lead writers to a stronger conclusion. The sample below demonstrates what a general outline looks like:

I. Introduction
    1. Background
    2. Thesis statement
II. Body
    1. Point A
        a. Supporting evidence
        b. Supporting evidence
    2. Point B
        a. Supporting evidence
        b. Supporting evidence
    3. Point C
        a. Supporting evidence
        b. Supporting evidence
III. Conclusion
    1. Restate main points of the paper.
    2. End with something memorable.

Drafting/Writing

Now it comes time to actually write the essay. In this stage, writers should follow the outline they developed in the brainstorming process and try to incorporate the useful sentences penned in the freewriting exercise. The main goal of this phase is to put all the thoughts together in cohesive sentences and paragraphs.

It is helpful for writers to remember that their work here does not have to be perfect. This process is often referred to as **drafting** because writers are just creating a rough draft of their work. Because of this, writers should avoid getting bogged down on the small details.

Referencing Sources

Anytime a writer quotes or paraphrases another text, they will need to include a citation. A **citation** is a short description of the work that a quote or information came from. The style manual your teacher wants you to follow will dictate exactly how to format that citation. For example, this is how one would cite a book according to the APA manual of style:

- Format: Last name, First initial, Middle initial. (Year Published) Book Title. City, State: Publisher.
- Example: Sampson, M. R. (1989). Diaries from an Alien Invasion. Springfield, IL: Campbell Press.

Revising

Revising offers an opportunity for writers to polish things up. Putting one's self in the reader's shoes and focusing on what the essay actually says helps writers identify problems—it's a movement from the mindset of writer to the mindset of editor. The goal is to have a clean, clear copy of the essay.

The main goal of the revision phase is to improve the essay's flow, cohesiveness, readability, and focus. For example, an essay will make a less persuasive argument if the various pieces of evidence are scattered and presented illogically or clouded with unnecessary thought. Therefore, writers should consider their essay's structure and organization, ensuring that there are smooth transitions between sentences and paragraphs. There should be a discernable introduction and conclusion as well, as these crucial components of an essay provide readers with a blueprint to follow.

Additionally, if the writer includes copious details that do little to enhance the argument, they may actually distract readers from focusing on the main ideas and detract from the strength of their work. The ultimate goal is to retain the purpose or focus of the essay and provide a reader-friendly experience. Because of this, writers often need to delete parts of their essay to improve its flow and focus. Removing sentences, entire paragraphs, or large chunks of writing can be one of the toughest parts of the writing process because it is difficult to part with work one has done. However, ultimately, these types of cuts can significantly improve one's essay.

Lastly, writers should consider their voice and word choice. The voice should be consistent throughout and maintain a balance between an authoritative and warm style, to both inform and engage readers. One way to alter voice is through word choice. Writers should consider changing weak verbs to stronger ones and selecting more precise language in areas where wording is vague. In some cases, it is useful to modify sentence beginnings or to combine or split up sentences to provide a more varied sentence structure.

Editing

Rather than focusing on content (as is the aim in the revising stage), the editing phase is all about the mechanics of the essay: the syntax, word choice, and grammar. This can be considered the proofreading stage. Successful editing is what sets apart a messy essay from a polished document.

The following areas should be considered when proofreading:

- Sentence fragments
- Awkward sentence structure
- Run-on sentences
- Incorrect word choice
- Grammatical agreement errors
- Spelling errors
- Punctuation errors
- Capitalization errors

One of the most effective ways of identifying grammatical errors, awkward phrases, or unclear sentences is to read the essay out loud. Listening to one's own work can help move the writer from simply the author to the reader.

During the editing phase, it's also important to ensure the essay follows the correct formatting and citation rules as dictated by the assignment.

Recursive Writing Process

While the writing process may have specific steps, the good news is that the process is recursive, meaning the steps need not be completed in a particular order. Many writers find that they complete steps at the same time such as drafting and revising, where the writing and rearranging of ideas occur simultaneously or in very close order. Similarly, a writer may find that a particular section of a draft needs more development, and will go back to the prewriting stage to generate new ideas. The steps can be repeated at any time, and the more these steps of the recursive writing process are employed, the better the final product will be.

Practice Makes Prepared Writers

Like any other useful skill, writing only improves with practice. While writing may come more easily to some than others, it is still a skill to be honed and improved. Regardless of a person's natural abilities, there is always room for growth in writing. Practicing the basic skills of writing can aid in preparations for the SSAT.

One way to build vocabulary and enhance exposure to the written word is through reading. This can be through reading books, but reading any materials such as newspapers, magazines, and even social media counts towards practice with the written word. This also helps to enhance critical reading and thinking skills, through analysis of the ideas and concepts read. Think of each new reading experience as a chance to sharpen these skills.

Developing a Well-Organized Paragraph

A **paragraph** is a series of connected and related sentences addressing one topic. Writing good paragraphs benefits writers by helping them to stay on target while drafting and revising their work. It benefits readers by helping them to follow the writing more easily. Regardless of how brilliant their ideas may be, writers who do not present them in organized ways will fail to engage readers—and fail to accomplish their writing goals. A fundamental rule for paragraphing is to confine each paragraph to a single idea. When writers find themselves transitioning to a new idea, they should start a new paragraph. However, a paragraph can include several pieces of evidence supporting its single idea; and it can include several points if they are all related to the overall paragraph topic. When writers find each point becoming lengthy, they may choose instead to devote a separate paragraph to every point and elaborate upon each more fully.

An effective paragraph should have these elements:

- Unity: One major discussion point or focus should occupy the whole paragraph from beginning to end.

- Coherence: For readers to understand a paragraph, it must be coherent. Two components of coherence are logical and verbal bridges. In logical bridges, the writer may write consecutive sentences with parallel structure or carry an idea over across sentences. In verbal bridges, writers may repeat key words across sentences.

- A topic sentence: The paragraph should have a sentence that generally identifies the paragraph's thesis or main idea.

- Sufficient development: To develop a paragraph, writers can use the following techniques after stating their topic sentence:

- Define terms
- Cite data
- Use illustrations, anecdotes, and examples
- Evaluate causes and effects
- Analyze the topic
- Explain the topic using chronological order

A **topic sentence** identifies the main idea of the paragraph. Some are explicit, some implicit. The topic sentence can appear anywhere in the paragraph. However, many experts advise beginning writers to place each paragraph topic sentence at or near the beginning of its paragraph to ensure that their readers understand what the topic of each paragraph is. Even without having written an explicit topic sentence, the writer should still be able to summarize readily what subject matter each paragraph addresses. The writer must then fully develop the topic that is introduced or identified in the topic sentence. Depending on what the writer's purpose is, they may use different methods for developing each paragraph.

Two main steps in the process of organizing paragraphs and essays should both be completed after determining the writing's main point, while the writer is planning or outlining the work. The initial step is to give an order to the topics addressed in each paragraph. Writers must have logical reasons for putting one paragraph first, another second, etc. The second step is to sequence the sentences in each paragraph. As with the first step, writers must have logical reasons for the order of sentences. Sometimes the work's main point obviously indicates a specific order.

Topic Sentences

To be effective, a topic sentence should be concise so that readers get its point without losing the meaning among too many words. As an example, in *Only Yesterday: An Informal History of the 1920s* (1931), author Frederick Lewis Allen's topic sentence introduces his paragraph describing the 1929 stock market crash: "The Bull Market was dead." This example illustrates the criteria of conciseness and brevity. It is also a strong sentence, expressed clearly and unambiguously. The topic sentence also introduces the paragraph, alerting the reader's attention to the main idea of the paragraph and the subject matter that follows the topic sentence.

Experts often recommend opening a paragraph with the topic sentences to enable the reader to realize the main point of the paragraph immediately. Application letters for jobs and university admissions also benefit from opening with topic sentences. However, positioning the topic sentence at the end of a paragraph is more logical when the paragraph identifies a number of specific details that accumulate evidence and then culminates with a generalization. While paragraphs with extremely obvious main ideas need no topic sentences, more often—and particularly for students learning to write—the topic sentence is the most important sentence in the paragraph. It not only communicates the main idea quickly to readers; it also helps writers produce and control information.

Writing Prompt

On the Upper Level SSAT, you will be presented with two prompts—a creative story prompt and a traditional essay prompt. On the exam, you are to choose one of the two prompts and write your story or essay from that prompt.

Creative Essay

Write a creative essay that starts with the sentence below:

> She took a long gulp out of her canteen, and then started up the mountain.

Traditional Essay

Write an argumentative essay. Take a side on the following question:

> Students should be required to learn financial skills while in high school. Do you agree or disagree? Support your argument.